How & Why We Subnet
Lab Workbook

CertificationKits.com How & Why We Subnet Workbook

Printed in the United States of America

Second Revision, First Printing October 2014

Library of Congress Cataloging-in-Publication Number:

ISBN: 978-1-60458-796-8

Warning and Disclaimer
This book is designed to provide self-study labs to help you prepare for your Cisco CCNA certification exam. Every effort has been made to make this book as complete and accurate as possible, but no warranty or fitness is implied.

Trademark Acknowledgements
All terms mentioned in this book that are known to be trademarks or service marks have been appropriately capitalized. The publisher, author or CertificationKits LLC cannot attest to the accuracy of this information. Use of a term in this book should not be regarded as affecting the validity of any trademark or service mark.

Contents

Chapter 4
Meeting the Stated Design Requirements 68

Chapter 5
Finite Address Space and NAT 73

Answer Key

How to Use This Workbook

There are many publications out there that will just focus on how to do subnetting to pass your Cisco exam. But they really do not explain, with real world scenarios, why we subnet. There are also many other publications out there that explain subnetting in a 15 page chapter. I am sure you know from experience that a simple 15 page chapter does not cover subnetting in depth. It is just the same old information regurgitated by various authors. So, we feel you are going to be pleased with what we have put together for you. Instead of just telling you how to convert to binary, decimal, and find the number of subnets and valid hosts, we are going to tie all these concepts together with real world examples and lots and lots of practice.

By the time you are done reading this will you have had enough practice to confidently walk into your Cisco certification exam and ace every subnetting question. Additionally, you will really understand in depth why we subnet and how the data travels between nodes at the packet level. So, this is not a quick read. Take your time and go through all the review questions and make sure you understand them fully.

Without further ado, we are going to jump right in and start off detailing base concepts of IP addresses, IP address classes and default subnet masks. As we progress in the chapter, we will start to discuss VLSM (variable length subnet masks) so you can see how to alter subnet masks to meet the requirements of a specific number of hosts, subnets and a combination of hosts and subnets. Some of these items can be explained using more than one method. In those cases, we may show you a method that is purely mathematical such as x to the 2nd power or an alternate method using a chart as a visual aid so you can "see" it. We will do this all while also providing you the real world examples and practice questions you need to really understand subnetting!

Chapter 1

What Are IP Addresses & Why Do We Need Them?

The Internet

Originally, the goal of the Internet was to enable government branches, facilities and universities to connect to each other and share information and research. It has come a long way since then as nearly every modern home in the world has Internet connectivity.

Consider the fact that all computers connected to the Internet need to send and receive data. How does that work? In the current implementation, the TCP/IP suite of protocols provides end-to-end connectivity between hosts on the Internet. However, that presents us with another problem. How do we uniquely identify each and every host connected to the Internet to insure that data is sent and received by the correct host?

IP Addresses

To correctly identify hosts on the Internet, IP addresses were devised. Each host is assigned an address that uniquely identifies the host on the Internet for communication.

Figure 1.1: Two hosts connected to the Internet

Consider Figure 1.1 above, Hosts A and B are both connected to the Internet. Assume that Host A wants to send Host B data. Since each host on the Internet has its own IP address to uniquely identify itself, Host A can specify the destination IP address it wants the data sent to as an end point. Another component of the information that will be sent in the data packet is the source IP address the packet was sent from. Figure 1.2, demonstrates this concept.

Figure 1.2: Source and destination addressing

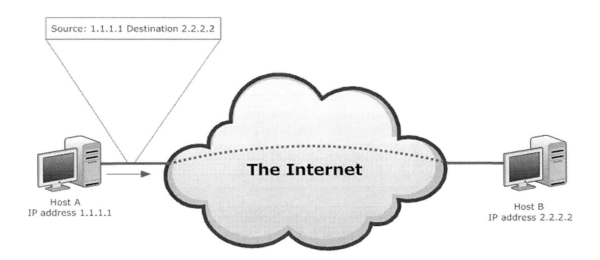

Now that we have established what IP addresses are and what their purpose is, how do we keep the IP addresses organized so that no two hosts get assigned the same IP address? The Internet Assigned Numbers Authority (IANA) was created to allocate and keep track of the IP addresses assigned. They in turn delegate allocations of IP addresses to regional Internet registries (RIRs). The RIRs then divide their allocated address pools into smaller blocks and delegate them to Internet service providers and other organizations in their respective regions.

Format of IP Addresses

IP addresses are 32 bit values. These 32 bits (32 bits equals 4 bytes) are broken up into four octets which are 8 bits each (8 bits equals 1 byte). The IP address is represented in the form of a.b.c.d where a, b, c and d are numbers in the range 0 – 255. While they are represented in regular decimal digits commonly referred to as the dotted decimal notation for human convenience, they are actually binary or base 2 and are seen by the IP devices as such.

This brings up the question of what is binary? You have probably heard of binary and bits before, which are represented using 1s and 0s. As mentioned above, binary is base 2, meaning it is comprised of 2 digits (1, 0), in comparison to decimal, which is base 10 and consists of 10 digits (0, 1, 2, 3, 4, 5, 6, 7, 8, 9). This means for every place or character in a decimal number there is a possibility of 10 digits. However, in binary you only have two possibilities; the value of a single digit is either 0 or 1. We will see some examples of this shortly.

To summarize, an IP address is represented in decimal for human convenience and actual network operations are handled in binary. Thus, we will need to be able to proficiently convert from binary to decimal and from decimal to binary to really understand what is happening at the IP level.

First we need to familiarize ourselves with the basic operation of decimal and binary, which is discussed in the next section.

Understanding Binary

Converting from Binary to Decimal

So how do we represent base 2 or binary numbers? It is quite simple. We will use the chart below in Figure 1.3 to help us explain it. The bits start from the right at a value of 1, and the value doubles as we move towards the left. Each bit has a corresponding value depending on where it resides in our chart. If the bit has a value of 1, it is considered on and that value will be included in our total. If the bit has a value of 0, it is considered off and it is not included in our total.

For example, consider the binary number 1101. Let us start from the rightmost bit, which is a 1. Since it is the rightmost bit, it represents a value of 1 and signifies that it is (on). The next bit towards the left is a 0, which has a value double to that of the first bit, 2. Notice that it is a 0 (off) however, meaning that it does not factor in the total value. The third bit is a 1 (on) and again represents double the value of the previous bit, 4. The fourth and final bit is double the third bit; it has a value of 8.

Figure 1.3: Binary Chart

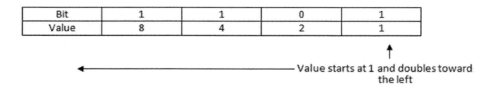

Bit	1	1	0	1
Value	8	4	2	1

Value starts at 1 and doubles toward the left

Once we have figured out which bits represent what values, we add the values represented of the on bits (1s) to find out the decimal representation of the binary bits.

values of on (1) bits are added to find the total value

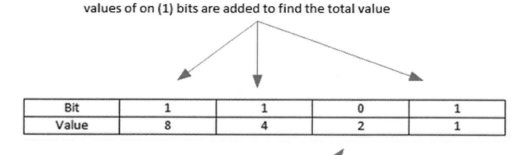

Bit	1	1	0	1
Value	8	4	2	1

An off (0) bit's value is not included in the sum

Adding the values of the on bits is the next step.

Bit	1	1	0	1
Value	8	4	2	1

$$8 + 4 + 1 = 13$$

Therefore, binary string 1101 is equal to 13 in decimal. As you have seen in the demonstration above, it is a simple process. Let us consider another example to familiarize ourselves with the process.

Convert the binary string 01001101 to decimal.

First we build a blank chart similar to the one we used above with the values starting at one on the right and doubling towards the left.

Figure 1.4: Blank Binary Chart

Bit								
Value	128	64	32	16	8	4	2	1

Then we will fill in the binary numbers in the chart as shown in Figure 1.5. Again, values of on bits (1) are added to find the total value and values of off (0) bits are ignored. Then we will add up the value of all the on bits to reveal the decimal value of our binary number as shown in Figure 1.5.

Figure 1.5: Binary Chart

Bit	0	1	0	0	1	1	0	1
Value	128	64	32	16	8	4	2	1

$$0 + 64 + 0 + 0 + 8 + 4 + 0 + 1 = 77$$

Thus the binary string 01001101 has a value of 77. In this case, note that the string started with a 0, 0s in string beginnings are called "leading zeros." Leading zeros have no significance whatsoever on the actual value and could be omitted; for example, 000010011 is the same as 10011. However, they usually listed due to a syntax requirement, such as in IP addresses. IP addresses are composed of 32 bits divided into 4 octets (an octet is an 8 bit string). Hence, if there are leading zeros, they should be included to have a complete 32 bit address. For example; assume we have a 32 bit IP address with the string 00101100 00001101 11010100 00101101. We can omit the leading zeros and represent it as 101100 1101 11010100 101101, but that may not be recognized properly.

Converting from Decimal to Binary

The process of converting decimal to binary is a little different from converting binary to decimal. Instead of having the bits already in the columns and summing up their values to get the decimal number; we start off with a decimal number and determine which bits are 1s and which are 0s in our binary column chart. We will do this by placing a 1 (or as some may say, turning on the bit as IP devices see bits represented by 1s as on and bits represented by 0s as off) in the highest binary column that can be turned on while subtracting the column value from the decimal number we are converting without creating a negative remainder. We will repeat this process until we have a decimal remainder of zero. I know that may sound a little confusing, but I am sure a few examples will clarify it for you and in no time you will find the process quite simple.

For clarification, consider this example.

Convert 52 to binary. We will use the same table.

Bit								
Value	128	64	32	16	8	4	2	1

Since we need to find out what the binary string is, we will fill in the bits fields following the procedure demonstrated previously.

Can we subtract 128 from 52 without resulting in a number less than zero (negative number)? No. That would not be possible, because we will get -76. So put a 0 bit in the 128 value and try the next value.

Bit	0							
Value	128	64	32	16	8	4	2	1

Can we subtract 64 from 52 without resulting in a number less than zero? No. That would not be possible, because we will get -12. So put a 0 bit in the 64 value and try the next value.

Bit	0	0						
Value	128	64	32	16	8	4	2	1

The next value is 32; can we subtract 32 from 52 without resulting in a number less than 0? Yes, we would have 20 left, so we put a 1 bit in the 32 value and find the next value that we can subtract from 20 without resulting in a number less than zero.

Bit	0	0	1					
Value	128	64	32	16	8	4	2	1

We now have 32 of 52, we need to allocate the remaining 20. Can we subtract 16 from 20? Yes we can, and we would have a remainder of 4. Put a 1 in the 16 value bit field and find the next value we can subtract 4 from without resulting in a number less than 0.

Bit	0	0	1	1				
Value	128	64	32	16	8	4	2	1

We still have 4 left. Can we subtract 8 from 4 without resulting in a number less than zero? No we cannot, that would result in a -4. So we put a 0 in the 8 value bit field and move on to the next one.

Bit	0	0	1	1	0			
Value	128	64	32	16	8	4	2	1

Can we subtract 4 from 4 without resulting in a number less than 0? Yes we can! We would be left with a 0, which is exactly what we want the result to be! Put a 1 in the 4 field.

Bit	0	0	1	1	0	1		
Value	128	64	32	16	8	4	2	1

Now that we have a 0; can we subtract 2 from 0 without having a number less than zero? How about 1 from 0? We cannot do either, as both would result in a number less than 0. So we put 0s in their respective value fields.

Bit	0	0	1	1	0	1	0	0
Value	128	64	32	16	8	4	2	1

Therefore, 52 is 00110100 in binary. Following the leading zeros rule, we can drop them and have an answer of 110100. Note that both are considered correct, one is just a little shorter and some may say easier to read.

As previously mentioned, IP addresses are divided into 4 sets of 8 bits known as octets (Note that 8 bits or an octet is also equal to a byte, 1 byte = 8 bits) which can range from 0-255. The tables of bit values used above demonstrate this; they make it clear why an octet is limited to a maximum total value of 255. If we turn on all the bits (set them to 1) and sum up all the values; 128 + 64 + 32 + 16 + 8 + 4 + 2 + 1 the sum will be equal to 255.

Exam Tip – Currently when you take the exam you are provided 90 minutes to complete it. Trust me, you will need every one of those precious minutes. So how can you afford yourself a few extra minutes? Well the exam does not start until you click the Start Exam button. Thus draw out all of your charts on the laminated boards provided to you by the testing center before you click the Start Exam button. This just might give you the extra 5 minutes you need to finish all the questions in the allotted time. Always ensure you double check your chart as if it is wrong you may get all your subnetting questions wrong!

IP Address Conversion - Decimal to Binary

To convert IP addresses to binary we will now take the decimal numbers and convert to binary. For example, let us convert the IP address 12.89.176.155 to binary.

To see how simple it is, we will tackle it octet by octet. Since each octet has a maximum of 8 bits, we should have fields for 8 bits and their corresponding values.

Bit								
Value	128	64	32	16	8	4	2	1

Let us consider the first octet, 12. Can we subtract 128 from 12? No? How about 64 from 12? Or 32 from 12? Or even 16 from 12? We can't subtract any of those from 12 without getting a number less than zero, so we put 0s in the respective value's bit field and keep trying to find the biggest value we can subtract 12 from, without having a result less than zero.

Bit	0	0	0	0				
Value	128	64	32	16	8	4	2	1

Can we subtract 8 from 12? Of course! That leaves us with 4. So we put a 1 in the 8 value bit field. Now can we subtract 4 from 4 without having a result less than zero? Yes! Once we have 0 left, we can just enter 0s in the remaining fields towards the right as we can't subtract anything from zero without resulting in a number less than zero.

Bit	0	0	0	0	1	1	0	0
Value	128	64	32	16	8	4	2	1

Great! So 12 in binary is 00001100, with 4 leading zeros; which can be eliminated based on the leading zeros rule but might cause confusion because an octet is expected to have 8 bits.

The next octet is 89. We follow the same steps to convert it to binary.

Bit								
Value	128	64	32	16	8	4	2	1

Again we find the largest value we can subtract from 89 without resulting in a number less than zero. Can we subtract 128 from 89? We cannot, so a 0 goes into that bit field. How about 64? Yes we can, and that leaves us with 25. Put a 1 in the 64 value bit field.

Bit	0	1						
Value	128	64	32	16	8	4	2	1

We are left with 25; can we subtract 32 from 25? No, so a 0 goes into that field. How about 16 from 25? Yes, so a 1 goes into that field leaving us with 9.

Bit	0	1	0	1				
Value	128	64	32	16	8	4	2	1

Now we have to find the largest value towards the right that we can subtract from 9 without resulting in a number less than 0. Can we subtract 8 from 9? Yes we can, which leaves us with 1. So a 1 goes into that value's bit field.

Bit	0	1	0	1	1			
Value	128	64	32	16	8	4	2	1

Now what number can we subtract from 1 without having an answer less than 0? Well 4 and 2 both don't meet the requirements, so zeros go into those values' respective bit fields. We are left with 1 itself! Can we subtract 1 from 1 without resulting in a result less than 0? Of course!

Bit	0	1	0	1	1	0	0	1
Value	128	64	32	16	8	4	2	1

So 89 is 01011001 in binary. Great, let us convert the two remaining octets, 176 and 155.

You should be getting the hang of it by now; the steps in converting decimal to binary will be summarized for the last two octets.

Can 128 be subtracted from 176? Yes? Good! That leaves us with 48; a 1 bit goes into the 128 value bit field. Can we subtract 64 from 48? No we cannot, so a 0 goes into the 64 value bit field. How about 32 from 48? Yes we can. That leaves us with 16, so a 1 goes into the 32 value bit field. Next, can we subtract 16 from 16? Yes! That leaves us with a remainder of 0. So we enter 1 in the 16 value bit field and 0s in the subsequent bit fields.

Bit	1	0	1	1	0	0	0	0
Value	128	64	32	16	8	4	2	1

The Last octet is 155, same steps as all of the previous examples. Can we subtract 128 from 155? Yes we can, it leaves us with 27. So a 1 goes into the 128 value bit field. How about 64 from 27? Or 32 from 27? We cannot subtract either of those from 27 without resulting in a number less than 0, so 0s go into their value bit fields. How about 16? Yes, we can subtract 16 from 27. That leaves us with 11, so a 1 goes into the 16 value bit field. Can we subtract 8 from 11? Yes that leaves us with a remainder of 3, so a 1 goes there as well. Is it possible to subtract 4 from 3? No it is not, so a 0 goes in that value bit field. 2 from 3? Yes! That leaves us with 1, so a 1 goes in the 2 value bit field. Finally, we subtract 1 from 1 yielding a remainder of 0 and also putting a 1 in the 1 value bit field.

Bit	1	0	0	1	1	0	1	1
Value	128	64	32	16	8	4	2	1

As you have seen, we basically put a 1 in the field if we can subtract the value from the number while meeting the condition of not having a result less than 0. If it doesn't meet the condition, we simply put a 0 in the value bit field and try the next value.

**Exam Tip – When you are performing your decimal to binary conversions, your remainder at the end of the process should be 0. If you have any other remainder, you made a mistake and should redo the question.

IP Address Conversion - Binary to Decimal

Our example will be to convert binary strings 01011000. 10110010. 00101101. 10101001 to decimal. All we have to do is fill in the bits in the respective octet fields. Again, to avoid getting overwhelmed, we will tackle this octet by octet.

Bits	Eighth (leftmost)bit	Seventh bit	Sixth bit	Fifth bit	Fourth bit	Third bit	Second bit	First(rightmost) bit
1st octet	0	1	0	1	1	0	0	0
Value	128	64	32	16	8	4	2	1

All we have to do is add all of the values with 1 bits set. Which are 64, 16 and 8, 64 + 16+ 8 = 88; giving this octet a total value of 88. Giving us 88.b.c.d

Bits	Eighth (leftmost)bit	Seventh bit	Sixth bit	Fifth bit	Fourth bit	Third bit	Second bit	First(rightmost) bit
2nd octet	1	0	1	1	0	0	1	0
Value	128	64	32	16	8	4	2	1

Get the values of bits with 1s in the bit field and add them. 128 + 32 + 16 + 2 = 178, so it is 88.178.c.d

Bits	Eighth (leftmost)bit	Seventh bit	Sixth bit	Fifth bit	Fourth bit	Third bit	Second bit	First(rightmost) bit
3rd octet	0	0	1	0	1	1	0	1
Value	128	64	32	16	8	4	2	1

Repeat the same steps, 32 + 8 + 4 + 1 = 45. Making it 88.178.45.d

Bits	Eighth (leftmost)bit	Seventh bit	Sixth bit	Fifth bit	Fourth bit	Third bit	Second bit	First(rightmost) bit
4th octet	1	0	1	0	1	0	0	1
Value	128	64	32	16	8	4	2	1

Last octet! 128 + 32 + 8 + 1 = 169, yielding a final answer of 88.178.45.169.

Basically it works the same way, just more to do.

Practice Tip – you can download our CCNA Subnet Calculator for some additional practice at http://www.CertificationKits.com/cisco-ccna-subnet-calculator/ and also check your answers with it.

Exam Tip – If during your conversion process you resulting answer is a 0 or 255, it does not mean your answer is necessarily incorrect. Remember that they are both valid numbers. A few examples would be 192.168.0.1 or 192.168.1.255.

Exercise 1.1: Binary to Decimal Practice

Convert the following digits from binary to decimal.

1) 11100111.00110010.00101101.11111011

Bits	Eighth (leftmost)	Seventh bit	Sixth bit	Fifth bit	Fourth bit	Third bit	Second bit	First bit	**Total value**
1st octet									
2nd octet									
3rd octet									
4th octet									
Value	128	64	32	16	8	4	2	1	

The decimal IP address is _____.

2) 00001110.01101110.00011001.00101010

Bits	Eighth (leftmost)	Seventh bit	Sixth bit	Fifth bit	Fourth bit	Third bit	Second bit	First bit	**Total value**
1st octet									
2nd octet									
3rd octet									
4th octet									
Value	128	64	32	16	8	4	2	1	

The decimal IP address is _____.

3) 11011101.00000001.11011101.01110110

Bits	Eighth (leftmost)	Seventh bit	Sixth bit	Fifth bit	Fourth bit	Third bit	Second bit	First bit	Total value
1st octet									
2nd octet									
3rd octet									
4th octet									
Value	128	64	32	16	8	4	2	1	

The decimal IP address is _____

4) 01010000.11111101.11110100.00010110

Bits	Eighth (leftmost)	Seventh bit	Sixth bit	Fifth bit	Fourth bit	Third bit	Second bit	First bit	Total value
1st octet									
2nd octet									
3rd octet									
4th octet									
Value	128	64	32	16	8	4	2	1	

The decimal IP address is _____.

5) 00000011.11111001.00110011.00111101

Bits	Eighth (leftmost)	Seventh bit	Sixth bit	Fifth bit	Fourth bit	Third bit	Second bit	First bit	Total value
1st octet									
2nd octet									
3rd octet									
4th octet									
Value	128	64	32	16	8	4	2	1	

The decimal IP address is _____.

6) 11000110.01011110.01111111.11111110

Bits	Eighth (leftmost)	Seventh bit	Sixth bit	Fifth bit	Fourth bit	Third bit	Second bit	First bit	**Total value**
1st octet									
2nd octet									
3rd octet									
4th octet									
Value	128	64	32	16	8	4	2	1	

The decimal IP address is _____.

7) 11111000.00000000.00000010.01100101

Bits	Eighth (leftmost)	Seventh bit	Sixth bit	Fifth bit	Fourth bit	Third bit	Second bit	First bit	**Total value**
1st octet									
2nd octet									
3rd octet									
4th octet									
Value	128	64	32	16	8	4	2	1	

The decimal IP address is _____.

8) 01000101.11111111.01011100.01111110

Bits	Eighth (leftmost)	Seventh bit	Sixth bit	Fifth bit	Fourth bit	Third bit	Second bit	First bit	**Total value**
1st octet									
2nd octet									
3rd octet									
4th octet									
Value	128	64	32	16	8	4	2	1	

The decimal IP address is _____.

9) 00000100.00001000.00000100.11111010

Bits	Eighth (leftmost)	Seventh bit	Sixth bit	Fifth bit	Fourth bit	Third bit	Second bit	First bit	**Total value**
1st octet									
2nd octet									
3rd octet									
4th octet									
Value	128	64	32	16	8	4	2	1	

The decimal IP address is _____.

10) 11011111.01110011.01000001.00101011

Bits	Eighth (leftmost)	Seventh bit	Sixth bit	Fifth bit	Fourth bit	Third bit	Second bit	First bit	**Total value**
1st octet									
2nd octet									
3rd octet									
4th octet									
Value	128	64	32	16	8	4	2	1	

The decimal IP address is _____.

Exercise 1.2 Decimal to Binary Practice

Convert the following decimal numbers to binary.

1) 231.50.45.251

Bits	Eighth (leftmost)	Seventh bit	Sixth bit	Fifth bit	Fourth bit	Third bit	Second bit	First bit	Total value
1st octet									
2nd octet									
3rd octet									
4th octet									
Value	128	64	32	16	8	4	2	1	

The binary representation is _____.

2) 14.110.25.42

Bits	Eighth (leftmost)	Seventh bit	Sixth bit	Fifth bit	Fourth bit	Third bit	Second bit	First bit	Total value
1st octet									
2nd octet									
3rd octet									
4th octet									
Value	128	64	32	16	8	4	2	1	

The binary representation is _____.

3) 221.1.221.118

Bits	Eighth (leftmost)	Seventh bit	Sixth bit	Fifth bit	Fourth bit	Third bit	Second bit	First bit	Total value
1st octet									
2nd octet									
3rd octet									
4th octet									
Value	128	64	32	16	8	4	2	1	

The binary representation is _____

4) 80.253.244.22

Bits	Eighth (leftmost)	Seventh bit	Sixth bit	Fifth bit	Fourth bit	Third bit	Second bit	First bit	Total value
1st octet									
2nd octet									
3rd octet									
4th octet									
Value	128	64	32	16	8	4	2	1	

The binary representation is _____.

5) 3.249.51.61

Bits	Eighth (leftmost)	Seventh bit	Sixth bit	Fifth bit	Fourth bit	Third bit	Second bit	First bit	Total value
1st octet									
2nd octet									
3rd octet									
4th octet									
Value	128	64	32	16	8	4	2	1	

The binary representation is _____.

6) 198.94.127.254

Bits	Eighth (leftmost)	Seventh bit	Sixth bit	Fifth bit	Fourth bit	Third bit	Second bit	First bit	Total value
1st octet									
2nd octet									
3rd octet									
4th octet									
Value	128	64	32	16	8	4	2	1	

The binary representation is _____.

7) 248.0.2.101

Bits	Eighth (leftmost)	Seventh bit	Sixth bit	Fifth bit	Fourth bit	Third bit	Second bit	First bit	Total value
1st octet									
2nd octet									
3rd octet									
4th octet									
Value	128	64	32	16	8	4	2	1	

The binary representation is _____.

8) 69.255.92.126

Bits	Eighth (leftmost)	Seventh bit	Sixth bit	Fifth bit	Fourth bit	Third bit	Second bit	First bit	Total value
1st octet									
2nd octet									
3rd octet									
4th octet									
Value	128	64	32	16	8	4	2	1	

The binary representation is _____.

9) 4.8.4.250

Bits	Eighth (leftmost)	Seventh bit	Sixth bit	Fifth bit	Fourth bit	Third bit	Second bit	First bit	**Total value**
1st octet									
2nd octet									
3rd octet									
4th octet									
Value	128	64	32	16	8	4	2	1	

The binary representation is _____.

10) 223.115.65.43

Bits	Eighth (leftmost)	Seventh bit	Sixth bit	Fifth bit	Fourth bit	Third bit	Second bit	First bit	**Total value**
1st octet									
2nd octet									
3rd octet									
4th octet									
Value	128	64	32	16	8	4	2	1	

The binary representation is _____.

Chapter 2

Subnets – What Are They & Why We Need Them?

Introduction

In this chapter we are going to slowly build up our knowledge as to what are subnets and why do we need them. We are going to do that by giving you a granular understanding of how hosts communicate and then it will become very clear to you why we use subnets to break up collision domains and control traffic. So bear with us as we go over some concepts that may be old hat for you while we definitely introduce some new concepts. Additionally, as we proceed you will start to see how the use of subnets helps conserve precious IP addresses which we officially ran out of in the IPv4 format. But don't worry, at the end of the book we will introduce some technologies that help mitigate those issues and I am quite sure you have heard of IPv6 which as the world coverts to it will provide us many, many more IP addresses for our disposal.

Media Access Control (MAC)

The media access control (MAC) addresses (also known as hardware addresses) are used to communicate at layer 2, the data link layer. The MAC address is burned-in the network card of the device that is connecting to the network, meaning that it is hard-coded. It is composed of 48 bits which are represented as 12 hexadecimal digits. These addresses are globally unique and are broken into two halves. The first half of the address represents the adapter manufacturer's ID assigned number, also called Orgainzational Unique Identifier or OID. Each manufacturer has a block of MAC addresses which it prefixes to the MAC address to identify its products. For instance, Cisco has multiple identifiers starting at 00-00-0C as well as others as they have purchased many companies over the years.

*To see a full list of organizations' unique identifiers refer to http://standards.ieee.org/develop/regauth/oui/oui.txt

The second half of the address is the individual adapter's identifier. MAC addresses are used to transport frames between devices in the same broadcast domain. A broadcast domain is simply the network segment in which broadcasts propagate to all the hosts on it (broadcast domains are broken up by routers or other layer 3 devices). In Figure 2.1, we have four computers from Sales attached to a Layer 2 switch.

Figure 2.1: Four hosts connected to the sales switch

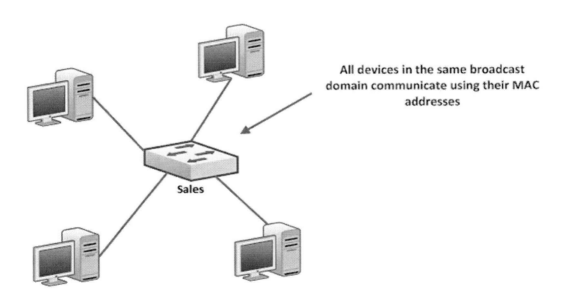

All devices in the same broadcast domain communicate using their MAC addresses

Sales

If devices are able to communicate using their MAC address, they are on the same local link (a local link refers to a local segment where hosts can communicate via Layer 2). This can be inefficient because of one simple layer 2 switching rule: all devices are in the same broadcast domain. This indicates it is a "flat network", if one host sends a broadcast all other devices on the network will receive it. Now imagine if there are a thousand or more hosts (a little known fact is that the recommended maximum number of hosts is 1024, on a 10Mbps network all bandwidth would be used by 1024 hosts just for contention and no actual data would be sent), all sending out broadcasts that all hosts will receive. This would create a lot of traffic. Most of this would be frivolous traffic which unnecessarily consumes network resources, bandwidth and computer processing power as each host will read each packet it receives. That is why we want to use unicast as opposed to broadcast to send our data. With unicast the data is only sent directly to the host we want to receive it. Thus no other host receives or processes the data.

You may wonder what the big deal is if I only have 4 computers on my network such as the example in Figure 2.1? Well what if we need to connect another department in our company to this one and have hundreds of nodes in each department? We can connect the Sales switch to the Marketing switch which also has nodes connected as shown in Figure 2.2 to set the stage.

Figure 2.2: Two Departments interconnected via Switches

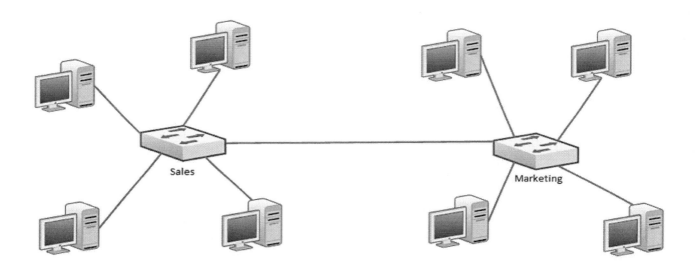

Before a device can communicate with another device on the local link, it has to know the destination's MAC address. MAC addresses are hard-coded to the network interface cards and cannot be changed unless spoofed.

How does a device learn the MAC address of another host? It resolves the destination IP address to the destination device's MAC address using ARP (Address Resolution Protocol). To demonstrate the operation of ARP, we are going to find out the MAC address of the computer with an IP address of 2.2.2.2. ARP sends out an ARP broadcast (to all hosts) asking who 2.2.2.2 is. Once the intended recipient (host 2.2.2.2) receives the message, it sends an ARP reply with the requested information (its MAC address) back to the source host. We will examine ARP packets in detail later on.

Imagine a host on the Sales network sends out an ARP broadcast request to a local server, observe figure 2.3.

Figure 2.3: ARP Broadcast Process

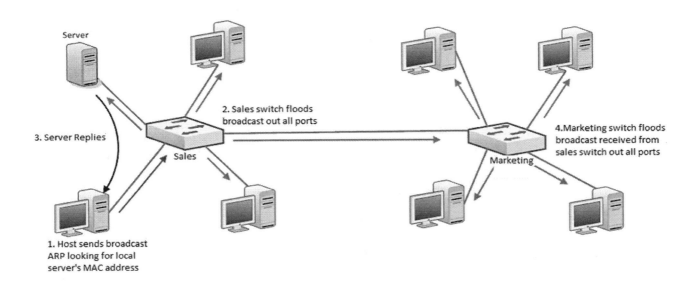

You can easily see that it quickly gets ugly when new devices are added to the network. The Marketing computers receive traffic that they should not. This makes it clear why MAC addresses are only used on local links and are not used to interconnect networks. They do not provide logical boundaries between networks or ways to identify logical grouping of hosts. Therefore, they cannot scale to large networks such as enterprise environments and the Internet.

So what does a MAC address look like? A MAC addresses as we mentioned is composed of 48-bits which are represented in hexadecimal and can be written in a few ways as shown below which are all considered correct:

ABCD.1234.5678

AB-CD-12-34-56-78

AB:CD:12:34:56:78

Go to a command prompt on your computer and do an "ipconfig /all" and look at the Physical Address field to see the MAC address of your computer as shown below in Figure 2.4.

Figure 2.4: MAC Address Example

```
Ethernet adapter Local Area Connection:

    Connection-specific DNS Suffix   .  : home
    Description  . . . . . . . . . . . : Atheros PCI-E Ethernet Controller
    Physical Address. . . . . . . . . : 90-E6-BA-0D-64-AF
    IPv4 Address. . . . . . . . . . . : 192.168.1.3(Preferred)
    Subnet Mask . . . . . . . . . . . : 255.255.255.0
    Default Gateway . . . . . . . . . : 192.168.1.1
    DHCP Server . . . . . . . . . . . : 192.168.1.1
    DNS Servers . . . . . . . . . . . : 192.168.1.1
```

Network IDs & Host IDs

As previously mentioned, IP addresses are the solution to our traffic segregation issues as they provide logical addressing by breaking an address into two portions; a Network ID and a Host ID as you can see in Figure 2.5.

Figure 2.5: IP Address Format

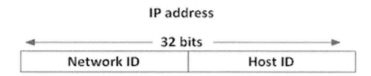

Basically, the Network ID portion is used to identify the network, while the Host ID is used to identify a host in that network. This provides hierarchical addressing and scalability. Devices belonging to the same network will all share the same "Network ID" but their "Host ID" should be unique to them.

Addresses were originally described as "classful", meaning that a certain set number of bits are used to represent the Network ID for each class. We have classes A, B and C with each defining a number of bits set to be used as the Network ID and Host ID. There are also classes D and E, which are reserved for multicast and experimental purposes respectively. However, classes D and E are not relevant to this discussion as these addresses cannot be assigned nor used in a unicast IP network. Let us first look at a quick chart that shows us all the IP Address Classes and some default information. We will then go into greater detail about the chart as we progress in our studies.

Figure 2.6: Address Class Chart

Class	Class A	Class B	Class C
1st Octet Valid Value	1-126*	128-191	192-223
Default Subnet Mask	255.0.0.0	255.255.0.0	255.255.255.0
Default # of Network Bits	8	16	24
Default # of Host Bits	24	16	8

* 127 is not listed as it is reserved for loopback.

For now, just take a quick glance at our chart in Figure 2.6 and notice that for each class the default number of network and host bits adds up to 32 bits. This is the exact same number of bits that comprises our IP address. Hopefully this makes sense to you at this point.

Class A address

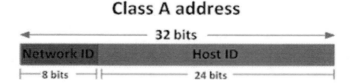

Class A addresses have 8 bits set for the network ID and the leftmost bit of the first octet (network ID) is always 0. Accordingly, that leaves 24 bits for the host ID. Per our table in Figure 2.7, if the leftmost bit which is in bold, is always 0 for a Class A address, then the maximum value of the first octet with the rest of the bits on is (64 + 32 + 16 + 8 + 4 + 2 +1) 127 (keep in mind, you can't really use the 127 network as it is a reserved range for loopback).

Figure 2.7: Binary Chart – Class A Maximum Value

Bits	Eighth (leftmost)bit	Seventh bit	Sixth bit	Fifth bit	Fourth bit	Third bit	Second bit	First(rightmost) bit
1st octet	**0**	1	1	1	1	1	1	1
Value	128	64	32	16	8	4	2	1

The minimal value would be 1 as we can't have a network address of 0. We showed this example below in Figure 2.8 in reverse due to the uniqueness of this one example having our starting value of 1.

Figure 2.8: Binary Chart – Class A Minimum Value

Bits	Eighth (leftmost)bit	Seventh bit	Sixth bit	Fifth bit	Fourth bit	Third bit	Second bit	First(rightmost) bit
1st octet	**0**	0	0	0	0	0	0	1
Value	128	64	32	16	8	4	2	1

Hence, class A addresses range in the first octet from 1 (lowest possible value bit) to 127 (highest possible bit value, but reserved for loopback). While the 2nd, 3rd and 4th octets can have any value. Making the range of addresses 1.0.0.0 – 127.255.255.255 (highest address possible with the leftmost bit of the network ID set to 0). You might wonder why we can't use addresses in the 0 range. Those in addition to a few more addresses are reserved. We will discuss them later.

But for now we also want to determine the number of networks in a Class A address mathematically. We can do that by calculating 2^7. Why is that? Well we have 0 as the most significant bit and that then leaves 7 bits left for the network number. Also note that since we have 24 host bits, we can assign 16,777,214 hosts to that network. This is derived by multiplying 2 by itself <number of host bits>. So to help you visualize this and the proceeding concept, we have provided Figure 2.9. If you recall our previous address chart, this may look familiar with two new rows added which are in bold to help us understand the number of hosts and networks per address class.

Exam Alert – Ok, you now may be saying...Hey, wait. I see in some books it says to figure out the number of subnets per class it is 2 to the power of x minus 2. Why are you and some other books not showing the minus 2? That does not seem consistent. Which is right? Actually both are right; depending on the situation. The way the concept was originally presented from Cisco was to always subtract two from the number of valid subnets to address the network and the broadcast subnets. But as features evolved, so did the exam. Now according to Cisco, you should only subtract 2 from the number of valid subnets if one of a few scenarios apply: A) A classful protocol is in use and for exam purposes this can only be RIPv1 as IGRP is no longer on the exam. B) The "no ip subnet-zero" command is in the exam question configuration presented to you. So if you do not see RIPv1 or "no ip subnet-zero", then do not subtract 2 from the number of valid subets (don't confuse this with the number of valid hosts as we still do it for that). We will follow this model for the remainder of this book.

Class	Class A	Class B	Class C
1st Octet Valid Value	1-126	128-191	192-223
Default Subnet Mask	255.0.0.0	255.255.0.0	255.255.255.0
Default # of Network Bits	8	16	24
Default # of Networks Per Class	**126**	**16,384 (2^{14})**	**2,097,152(2^{21})**
Default # of Host Bits	24	16	8
Default # of Hosts Per Class	**16,777,214 (2^{24}-2)**	**65,534 (2^{16}-2)**	**254 (2^8-2)**

So in this case, we simply multiply 2 by itself 24 times, yielding an answer of 16,777,216 – 2 = 16,777,214. We subtract two because of the "network address" and the "broadcast address". The network address which represents the network itself is represented with all the host bits set to 0. So for example, network 10.0.0.0 has a network ID field of 10 (1st octet) and unset host bits (0.0.0; all bits are off). This address cannot be assigned to a host because it is used to represent the network the hosts are in. Therefore, it is subtracted from the total number of assignable host addresses. The other address we subtract is the broadcast address. This address is used to send a packet to all hosts in a network (hence the term broadcast). This is in contrast with unicast packets, which is sent to a single host. A broadcast is represented with all the host bits set to 1. For instance, using our 10.0.0.0 example, setting the hosts bits on in all 3 host ID octets would result in a broadcast address of 10.255.255.255. Those are the two addresses we subtract to derive the range of assignable host addresses, so the valid range would be from 10.0.0.1 to 10.255.255.254.

Class B address

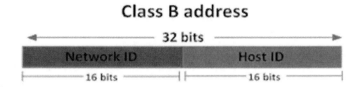

Class B addresses on the other hand have half the address (16 bits) for the network ID and the other half set aside for the host ID. So there are 16 network ID bits and 16 host ID bits. The two leftmost bits are 1 and 0 in that order for the network ID. As Figure 2.10 indicates below, the minimal value with all the remaining bits turned off is 128 as the string is 1000000 (eighth bit always set to 1 and the seventh bit to 0).

Figure 2.10: Binary Chart – Class B Minimum Value

Bits	Eighth (leftmost)bit	Seventh bit	Sixth bit	Fifth bit	Fourth bit	Third bit	Second bit	First(rightmost) bit
1st octet	**1**	**0**	0	0	0	0	0	0
Value	128	64	32	16	8	4	2	1

Accordingly, the maximum possible value with the eight bit on, the seventh bit off and the remaining bits on would be 128 + 32 + 16 + 8 + 4 + 2 + 1 (or simply 255 – 64) = 191 as shown in Figure 2.11.

Figure 2.11: Binary Chart – Class B Maximum Value

Bits	Eighth (leftmost)bit	Seventh bit	Sixth bit	Fifth bit	Fourth bit	Third bit	Second bit	First(rightmost) bit
1st octet	1	0	1	1	1	1	1	1
Value	128	64	32	16	8	4	2	1

Therefore, class B addresses range from 128 – 191. Next is an important concept to understand. Note that only the first octet is defined and the remaining octets can have any value; more specifically 128.0.0.0 – 191.255.255.255. The number of host bits is 16, so we multiply 2 by itself 16 times and subtract two, resulting in 65534 assignable host addresses per network ID. Let's also figure out the number of networks for a Class B address. In this case we have two leading bits, 1 and 0 and thus are left with 14 bits for the network number. 2^{14} equals 16,384 networks as shown back in Figure 2.9.

Class C address

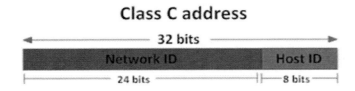

Class C addresses start with 110 as the leading bits. This provides 24 bits for the network ID and 8 bits for the host ID. Using our table in Figure 2.12 below, the minimal value with both the 128 and 64 value bits set on(128 + 64 = 192) is 192 with all the remaining bits off; making the range 192.0.0.0 – 223.255.255.255, again bits on the 2nd, 3rd and 4th octets can assume any value. This time around, we have 8 host bits, so we multiply 2 by itself 8 times and subtract 2 yielding 254 assignable host addresses. Let's also figure out the number of networks for a Class B address. In this case we have three leading bits, 100 and thus are left with 21 bits for the network number. 2^{21} equals 2,097,152 networks as shown back in Figure 2.9.

Figure 2.12: Binary Chart – Class C Minimum Value

Bits	Eighth (leftmost)bit	Seventh bit	Sixth bit	Fifth bit	Fourth bit	Third bit	Second bit	First(rightmost) bit
1st octet	1	1	0	0	0	0	0	0
Value	128	64	32	16	8	4	2	1

Finally we can see that the maximal value in Figure 2.13 below for the first octet in a class C address is 223 (simply 255 – 32, because if all bits are set then they would total 255. However, we do know that the 32 value bit is off, so we just subtract it from the total value).

Figure 2.13: Binary Chart – Class C Maximum Value

Bits	Eighth (leftmost)bit	Seventh bit	Sixth bit	Fifth bit	Fourth bit	Third bit	Second bit	First(rightmost) bit
1st octet	1	1	0	1	1	1	1	1
Value	128	64	32	16	8	4	2	1

Wait a minute, altogether classes A, B and C only range from 1.0.0.0 – 223.255.255.255. What about 224.0.0.0 – 255.255.255.255? These addresses go into classes D and E. Class D addresses range from 224.0.0.0 – 239.255.255.255 and are reserved for multicasts and class E addresses are reserved for future use and range from 240.0.0.0 – 254.255.255.255.   Thus they are not applicable to our discussion. Finally, 255.255.255.255 is reserved as a broadcast address.

How does all of this fit into our earlier example? A router will be inserted between our Sales and Marketing networks to prevent broadcasts from propagating between the two networks as shown in Figure 2.14. Now you can see, the networks are now logically separated by our layer 3 (IP) device, a router.

To cement the concept of finding the number of valid hosts, we are going to complete practice questions later on.

Figure 2.14: Two Departments Separated by a Router

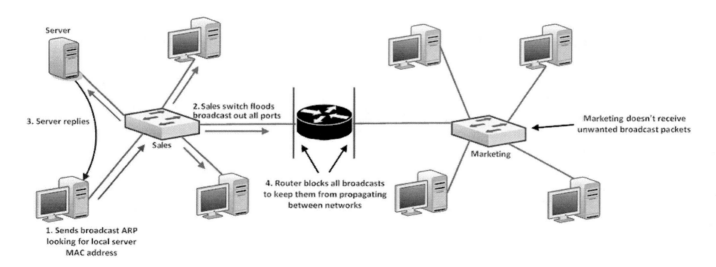

If there is one thing a router was invented to do other than connect networks, it is to block broadcasts. Basically, a router keeps track of which networks (via route tables) are reachable out which interfaces and forwards packets based on that information.

If communication between different networks is desired (which it will be in most cases), each network is assigned a unique network ID and each host within that network is assigned a host ID. I like to tell students to think of the network ID like your street address and a host ID as your house number.   You have lots of houses on your street (network ID) and the house number (host ID) is unique to just your house. Let us assume that the Sales network is 10.0.0.0 and the

Marketing network is 11.0.0.0; remember that the first octet is the network ID and the 3 remaining octets are used for the host ID portion in a class A network, 1.0.0.0 – 127.255.255.255 (in which both 10.0.0.0 and 11.0.0.0 are included). Let's review this scenario in Figure 2.15.

Figure 2.15: Router's Route Table

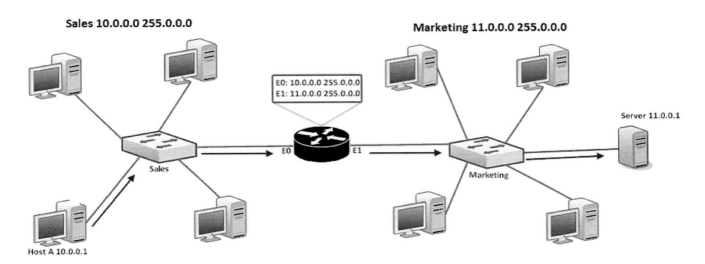

Host A (10.0.0.1), wants to communicate with the Server (11.0.0.1) in the Marketing network. The first thing that Host A does is apply the subnet mask to the IP address. A subnet mask allows you to identify which part of the IP address is reserved for the network and which part is available for the host. For example, a class A IP address of 10.0.0.0 has the first octet (all 8 bits) signifying the network ID, hence the subnet mask would reflect that as 255.0.0.0. Meaning that all bits in the first octet (with a total of 255 as previously mentioned) are part of the network ID and none of the other bits in any of the remaining octets is part of the network ID. Which is accurate, because only 8 bits in a class A network are used for the network ID portion while the remaining 24 bits (3 octets) are used for the host ID. In the same manner, since two octets in a class B address (16 bits) signify the network ID portion, the subnet mask would be 255.255.0.0 reflecting that all 16 bits of the first two octets represent the network ID for Class B addresses. Well what about class C addresses? Well we mentioned that 24 bits (3 octets) represent the network ID portion for Class C addresses. What subnet mask does that yield considering that all bits in an octet representing the network ID are added? Yes, 255.255.255.0 is the correct answer.

So as you can see from our examples above, you can probably visually look at an IP address such as 10.0.0.1 with a subnet mask of 255.0.0.0 and realize the 255(all bits on) in the first octet "masks" out the network ID. But let's take a very simple example like this and see how a computer actually sees it using ANDing and binary math.

First we need to convert the IP address and subnet mask to binary.

So how does binary math work? Simply 1+1=1. 1+0=0. 0+0=0. So starting from left to right you add the IP address bit to the Subnet mask bit and come up with the resultant Network ID as shown in Figure 2.16.

Figure 2.16: Binary ANDing

IP Address   00001010.00000000.00000000.00000001

Subnet mask   11111111.00000000.00000000.00000000

Network ID   **00001010.00000000.00000000.00000000**

So you can see the resultant network ID once you convert it from binary to decimal is 10.0.0.0 by doing binary math. The same as the answer we came up with above. So why take the long way and do all this math when I can see that visually? Well, just give it a little bit until we discuss variable length subnet masks which at first you will not be able to do visually, but after sufficient practice, you will be able to do them in your head too. But you have to understand the theory behind it first to be able to do so.

Continuing with our example, host A checks its IP address and applies its subnet mask to it; 10.0.0.1 255.0.0.0 to determine its network ID is 10.0.0.0 as shown above in Figure 2.16.

It then compares that to the destination IP address of 11.0.0.1 with a subnet mask of 255.0.0.0 to determine the resultant network ID for the Server IP address which is 11.0.0.0 as shown in Figure 2.17.

Figure 2.17: Binary ANDing

IP Address   00001011.00000000.00000000.00000001

Subnet mask   11111111.00000000.00000000.00000000

Network ID   **00001011.00000000.00000000.00000000**

Host A realizes that the destination is on a different network since the network IDs (10.0.0.0 and 11.0.0.0) do not match. Host A then forwards the packet to the router, which acts as a default gateway on the segment (meaning if you don't know where the destination network is, send it to the default gateway). The router performs a lookup in its route table and finds that the destination network 11.0.0.0(Marketing) is reachable via interface E1. Thus it sends the packet out E1 and once the packet is on the correct subnet, it will then resolve the IP to MAC address for the host 11.0.0.1 via ARP.

So how does this look at the packet level? Well if you review the examples below you can see an ARP packet and an ARP reply packet to see how the information is constructed and generated. First we will examine Figure 2.18 which is an ARP packet.

Figure 2.18: ARP Packet

Source	Destination	Protocol	Info
AskeyCom_de:d1:bc	Azurewav_6e:9e:c5	ARP	Who has 11.0.0.1? Tell 11.0.0.10

```
⊟ Address Resolution Protocol (request)
      Hardware type: Ethernet (0x0001)
      Protocol type: IP (0x0800)
      Hardware size: 6
      Protocol size: 4
      Opcode: request (0x0001)
      [Is gratuitous: False]
      Sender MAC address: AskeyCom_de:d1:bc (b4:82:fe:de:d1:bc)
      Sender IP address: 11.0.0.10 (11.0.0.10)
      Target MAC address: 00:00:00_00:00:00 (00:00:00_00:00:00)
      Target IP address: 11.0.0.1 (11.0.0.1)
```

As we examine the ARP packet, here are a few things to note. It tells us that the node (as this could be a router, workstation or some other device) that is sending out the request has an IP address of 11.0.0.10 which is layer 3 of the OSI model. As we move down the stack we note that the workstation has a MAC address of b4:82:fe:de:d1:bc. If you go to a command prompt on your PC and type "arp –a" you will see how your workstation maps the IP address to a MAC address in your ARP cache. So that tells us all the information about the source or about the sender. As this travels out on the local wire, who should receive it? Well the target IP is 11.0.0.1. In this case the workstation is trying to resolve to the local router or default gateway, notice how the target MAC address is all zeros.

Once the host at 11.0.0.1 receives the packet, it processes it and formulates a reply. Now take a look at the ARP Reply packet in Figure 2.19.

Figure 2.19: ARP Reply

Source	Destination	Protocol	Info
Azurewav_6e:9e:c5	AskeyCom_de:d1:bc	ARP	11.0.0.10 is at 00:25:d3:6e:9e:c5

```
⊟ Address Resolution Protocol (reply)
      Hardware type: Ethernet (0x0001)
      Protocol type: IP (0x0800)
      Hardware size: 6
      Protocol size: 4
      Opcode: reply (0x0002)
      [Is gratuitous: False]
      Sender MAC address: Azurewav_6e:9e:c5 (00:25:d3:6e:9e:c5)
      Sender IP address: 11.0.0.1 (11.0.0.1)
      Target MAC address: AskeyCom_de:d1:bc (b4:82:fe:de:d1:bc)
      Target IP address: 11.0.0.10 (11.0.0.10)
```

First notice that it states this is an ARP reply as opposed to the packet in 2.18 which was an ARP request. Also note how the targets and senders addressed have consequentially flip flopped. This should make sense as in our original example in Figure 2.18 the source was 11.0.0.10 and now it is 11.0.0.1. Also note how now the Target MAC address is populated as our new source knows the MAC address of the source device which it will be sending the data back to. In a way, it does not even need to know the destination IP address at this point as the communication is MAC to MAC since we now resolved our Layer 3 IP address to the corresponding Layer 2 MAC address for both the source and target devices.

Now to get you some practice ANDing to figure out the network address, we are going to do 3 examples of basic ANDing to determine a network ID.

Exercise 2.1: Determining the Network ID ANDing Practice

1) IP Address 21.153.24.169 and subnet mask 255.0.0.0

Bits	1st Octet	2nd Octet	3rd Octet	4th Octet
IP Address				
Subnet Mask				
Network ID				

The IP address 21.153.24.169 is on the _____ network.

2) IP Address 131.1.54.69 and subnet mask 255.255.0.0

Bits	1st Octet	2nd Octet	3rd Octet	4th Octet
IP Address				
Subnet Mask				
Network ID				

The IP address 131.1.54.69 is on the _____ network.

3) IP Address 239.244.1.16 and subnet mask 255.255.255.0

Bits	1st Octet	2nd Octet	3rd Octet	4th Octet
IP Address				
Subnet Mask				
Network ID				

The IP address 239.244.1.16 is on the _____ network.

Finding the Network Address

Now that you are just about ANDing experts, we are going to show you how to address another type of question you will see on the exam. You are sometimes asked to determine what subnet a specific IP address is on when you are given a particular subnet mask. So with all the knowledge we gained, we are going to use the same address space we have been using and throw in an advanced concept that will be covered in greater detail in the next section. But just hang in and it will come together as we want to keep our ANDing momentum!

First we will start off with a bit of review. If on the exam (or real world) we need to find the network in which a host resides when they provide us with an IP address and a subnet mask, we simply apply the subnet mask to the IP address and we should easily see it. Well, only if we were a computer! Since we are not computers, we will explain what a computer does to determine this so you fully understand the concept. For this example, let's say we have a host IP address of 10.70.0.1 with a subnet mask of 255.0.0.0.

We will first convert the IP address to binary as shown in Figure 2.20 by filling in our chart.

Figure 2.20: Decimal to Binary Conversion Chart

Bits	Eighth (leftmost)bit	Seventh bit	Sixth bit	Fifth bit	Fourth bit	Third bit	Second bit	First(rightmost) bit
1st octet	0	0	0	0	1	0	1	0
2nd octet	0	1	0	0	0	1	1	0
3rd octet	0	0	0	0	0	0	0	0
4th octet	0	0	0	0	0	0	0	1
Value	128	64	32	16	8	4	2	1

The IP address of 10.70.0.1 has a binary string of 00001010.01000110.00000000.00000001

Then we convert the subnet mask to binary as shown in Figure 2.21.

Figure 2.21: Decimal to Binary Conversion Chart

Bits	Eighth (leftmost)bit	Seventh bit	Sixth bit	Fifth bit	Fourth bit	Third bit	Second bit	First(rightmost) bit
1st octet	1	1	1	1	1	1	1	1
2nd octet	0	0	0	0	0	0	0	0
3rd octet	0	0	0	0	0	0	0	0
4th octet	0	0	0	0	0	0	0	0
Value	128	64	32	16	8	4	2	1

Subnet mask 255.0.0.0 has a binary string of 11111111.00000000.00000000.00000000. Now we can AND the subnet mask to the binary IP address to determine the network in which the IP address resides on. How do you do it? Well, binary math as mentioned before is a 1 and 1 = 1. 1 and 0 =0. 0 and 0 = 0. So you "AND" the IP address bit and the subnet bit to get the resultant network ID binary answer. Let's take a look at the example in Figure 2.22.

Figure 2.22: Binary Representation for ANDing

IP Address   00001010.01000110.00000000.00000001

Subnet mask   11111111.00000000.00000000.00000000

Network ID   **00001010.00000000.00000000.00000000**

So when we convert the above Network ID from binary to decimal (so humans can read it easier), we get 10.0.0.0. Thus the IP address of 10.70.0.1 with a subnet mask of 255.0.0.0 resides on the 10.0.0.0 network. You might be saying, yeah, that is nothing new. We already went over that.

Now that was an easy example. Now I am going to give you a more complex example that will really make sense once you read the next section. But I promise you that if you just hang in there, it will make sense! For this advanced example, let us say that we have the same host 10.70.0.1 but this time with a subnet mask of 255.192.0.0. You may be saying, I have only seen subnet masks with all 255s in it. Just go with it as we want to focus on the math for now.

We will first convert the IP address to binary as shown in Figure 2.23.

Figure 2.23: Decimal to Binary Conversion Chart

Bits	Eighth (leftmost)bit	Seventh bit	Sixth bit	Fifth bit	Fourth bit	Third bit	Second bit	First(rightmost) bit
1st octet	0	0	0	0	1	0	1	0
2nd octet	0	1	0	0	0	1	1	0
3rd octet	0	0	0	0	0	0	0	0
4th octet	0	0	0	0	0	0	0	1
Value	128	64	32	16	8	4	2	1

The IP address of 10.70.0.1 has a binary string of 00001010.01000110.00000000.00000001

Then we convert the subnet mask to binary as shown in Figure 2.24.

Figure 2.24: Decimal to Binary Conversion Chart

Bits	Eighth (leftmost)bit	Seventh bit	Sixth bit	Fifth bit	Fourth bit	Third bit	Second bit	First(rightmost) bit
1st octet	1	1	1	1	1	1	1	1
2nd octet	1	1	0	0	0	0	0	0
3rd octet	0	0	0	0	0	0	0	0
4th octet	0	0	0	0	0	0	0	0
Value	128	64	32	16	8	4	2	1

Subnet mask 255.192.0.0 has a binary string of 11111111.11000000.00000000.00000000. Now we can AND the subnet mask to the binary IP address to determine the network in which the IP address resides on. How do you do it? Well again, binary math is a 1 and 1 = 1. 1 and 0 = 0. 0 and 0 = 0. So you "AND" the IP address bit and the subnet bit to get the resultant network ID binary answer. Let's take a look at the example in Figure 2.25.

Figure 2.25: Binary Representation for ANDing

IP Address   00001010.01000110.00000000.00000001

Subnet mask   11111111.11000000.00000000.00000000

Network ID   **00001010.01000000.00000000.00000000**

So when we convert the above resultant Network ID from binary to decimal (so humans can read it easier), we get 10.64.0.0. Thus the IP address of 10.70.0.1 with a subnet mask of 255.192.0.0 resides on the 10.64.0.0 network.

So we will take a quick break here to do some practice examples to reinforce the concept. Below are 3 IP Addresses and subnet masks which you can AND to determine the subnet that each IP address resides on.

Exercise 2.2: Finding the Subnet the Address Resides On

1) IP Address 21.153.24.169 and subnet mask 255.240.0.0

Bits	1st Octet	2nd Octet	3rd Octet	4th Octet
IP Address				
Subnet Mask				
Network ID				

The IP address 21.153.24.169 is on the _____ network.

2) IP Address 131.1.54.69 and subnet mask 255.255.192.0

Bits	1st Octet	2nd Octet	3rd Octet	4th Octet
IP Address				
Subnet Mask				
Network ID				

The IP address 131.1.54.69 is on the _____ network.

3) IP Address 239.244.1.16 and subnet mask 255.255.255.192

Bits	1st Octet	2nd Octet	3rd Octet	4th Octet
IP Address				
Subnet Mask				
Network ID				

The IP address 239.244.1.16 is on the _____ network.

Chapter 3

VLSM – Subnetting Without Being Wasteful!

Calculating the Number of Valid Subnets

With newly acquired knowledge, you have decided to address your company's engineering, sales and marketing departments using the 10.0.0.0 network. However, broadcast traffic must not propagate between departments as that presents security issues and also makes the network inefficient. An example of a problem this might cause is server reachability. The engineering department's information should not be accessible by the sales and marketing people and vice versa.

So how should we approach this? The broadcast domains need to be broken up. Each network is a broadcast domain, and we only have a 10.0.0.0 network. How do we break it up into 3 networks? Subnetting! We can "borrow" bits from the host portion of the address and use them to create a subnet ID.

The number of bits you borrow from the host ID depends on the number of subnets you want to create. Using the same rules of binary, every bit you borrow doubles the number of subnets you can create. We will go into more detail on this with a chart shortly.

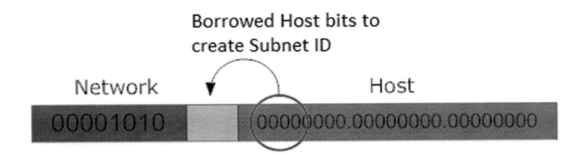

Subnetting is simple. All you have to do is figure out the desired number of subnets and borrow the corresponding number of bits you need from the host portion. For example, since an octet totals 255, we can create two subnets by dividing it by 128 which is the value of the first bit from the left (most significant). What if we need 3 subnets? We cannot divide 256 by 3 at the binary level, because none of the bit values (128, 64, 32, 16, 8, 4, 2, and 1) would result in an answer of 3. In that case, we can either create two subnets by dividing 256 by 128(which would not meet our requirement), or four subnets by dividing 256 by 64. Take a look at the chart below in Figure 3.1 and I think this will then make more sense.

Figure 3.1: Number of Subnets Chart

Bits	Eighth (leftmost)bit	Seventh bit	Sixth bit	Fifth bit	Fourth bit	Third bit	Second bit	First(rightmost) bit
Value	128	64	32	16	8	4	2	1
Number of subnets	2	4	8	16	32	64	128	256

As you can see, the number of subnets increases as the value of bits decreases, which is logical, because if we divide by a large number we would have a small answer. For example, 10 divided by 5 is 2. And if we divide by a small number, the answer would be a large number; 10 divided by 2 is 5.

The network ID is also known as a "prefix" because it is prefixed to all addresses in the network. In network 10.0.0.0, the network ID, 10 is the prefix. The subnet mask, which indicates which bits represent the network is also known as the "prefix length".

For instance, in class A network 10.0.0.0, the string of the first octet (00001010) is prefixed to all addresses in the network so that is the prefix length. Instead of representing this as 10.0.0.0 255.0.0.0 as this takes a lot of time to write out, we can also use something known as the CIDR (classless inter-domain routing) notation. Using this method, we simply count the number of bits that are on in the subnet mask and append that to the address. For our example above it would look like 10.0.0.0/8 as 255.0.0.0 only has the first 8 bits on.

So we are going to take a quick break from our example here to provide you an example of calculating the number of valid subnets for each address class.

How many valid subnets exist on the 19.0.0.0/13 network?

Masks	1st Octet	2nd Octet	3rd Octet	4th Octet
Class A mask	11111111	00000000	00000000	00000000
/13 Mask	11111111	**11111**000	00000000	00000000

Since it is a Class A (default of /8) we have 8 bits of the mask preset to act as network ID and an additional 5 bits which act as subnet bits (indicated in bold in the above table). The default 8 bits plus the 5 additional bits gives us our /13. As we already explained, to find the number of subnets we simply multiply 2 by itself <number of subnet bits> times, which in this case is 5; therefore, 2 multiplied by itself 5 times is 32. Hence, we can have 32 subnets from the 19.0.0.0/13 network.

Another way to look at it if you don't like the power of 2 is to refer back to the chart in Figure 3.1. If you look at the octet that has been subnetted, you could say it has the "excess mask bits". You can then apply them to the chart in Figure 3.1. Then simply look at the value for the least significant bit that is on. It lines up with the value that represents the number of subnets you have. Pretty neat, huh?

How many subnets are on the 131.5.0.0/18 network?

Masks	1st Octet	2nd Octet	3rd Octet	4th Octet
Class B mask	11111111	11111111	00000000	00000000
/18 Mask	11111111	11111111	**11**000000	00000000

131.5.0.0 /18 is a class B by default (16 network bits) this indicates that we have 2 additional bits for the subnet portion of the address. Again, we multiply 2 by itself <number of subnet bits> times, i.e. 2 multiplied by 2, which yields 4 subnets.

How subnets are there on the 240.224.1.0/27 network?

Masks	1st Octet	2nd Octet	3rd Octet	4th Octet
Class C mask	11111111	11111111	11111111	00000000
/27 Mask	11111111	11111111	11111111	**111**00000

Lastly, 240.224.1.0 /27, a class C address, (24 network bits) has 3 subnet bits. 2 times itself 3 times is 8. So we have 8 subnets.

Now to reinforce the concept, we are going to do three examples for each address class.

Exercise 3.1: Determining the Number of Valid Subnets for Class A Addresses Practice

1) 24.0.0.0 /14

Masks	1st Octet	2nd Octet	3rd Octet	4th Octet
Class A mask				
/14 Mask				

The number of valid subnets is _____.

2) 2.0.0.0 /10

Masks	1st Octet	2nd Octet	3rd Octet	4th Octet
Class A mask				
/10 Mask				

The number of valid subnets is _____.

3) 97.0.0.0 /17

Masks	1st Octet	2nd Octet	3rd Octet	4th Octet
Class A mask				
/17 Mask				

The number of valid subnets is _____.

Exercise 3.2: Determining the Number of Valid Subnets for Class B Addresses Practice

1) 142.42.0.0 /25

Masks	1st Octet	2nd Octet	3rd Octet	4th Octet
Class B mask				
/25 Mask				

The number of valid subnets is _____.

2) 169.1.0.0 /19

Masks	1st Octet	2nd Octet	3rd Octet	4th Octet
Class B mask				
/19 Mask				

The number of valid subnets is _____.

3) 131.31.0.0 /20

Masks	1st Octet	2nd Octet	3rd Octet	4th Octet
Class B mask				
/20 Mask				

The number of valid subnets is _____.

Exercise 3.3: Determining the Number of Valid Subnets for Class C Addresses Practice

1) 223.21.1.0/30

Masks	1st Octet	2nd Octet	3rd Octet	4th Octet
Class C mask				
/30 Mask				

The number of valid subnets is _____.

2) 210.21.9.0/29

Masks	1st Octet	2nd Octet	3rd Octet	4th Octet
Class C mask				
/29 Mask				

The number of valid subnets is _____.

3) 253.1.1.0/26

Masks	1st Octet	2nd Octet	3rd Octet	4th Octet
Class C mask				
/26 Mask				

The number of valid subnets is _____.

Back to our example, let us do a quick review. We have decided to address our company's engineering, sales and marketing departments using the 10.0.0.0 network. However, broadcast traffic must not propagate between departments as that presents security issues and also makes the network inefficient. The engineering department's information should not be accessible by the sales and marketing people and vice versa.

So how should we approach this? The broadcast domains need to be broken up. Each network is a broadcast domain, and we only have a 10.0.0.0 network. How do we break it up into 3 networks? Subnetting!

Figure 3.2: 10.0.0.0 Address Space Subnetted

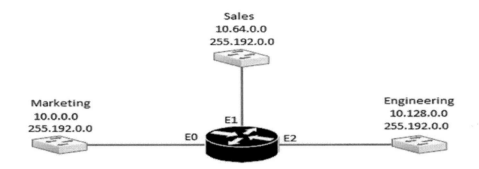

As you can see above in Figure 3.2, we have assigned the three address spaces from the 10.0.0.0 network and kept them separate by subnetting. We chose to create four subnets (which was the only logical choice) by dividing 256 by 64 which gives us 4 in the chart in Figure 3.3. That means to meet our goal of 4 subnets; we have to "steal" two bits from the host ID.

Figure 3.3: Number of Subnets Chart

Bits	Eighth (leftmost)bit	Seventh bit	Sixth bit	Fifth bit	Fourth bit	Third bit	Second bit	First(rightmost) bit
Value	128	64	32	16	8	4	2	1
No. of subnets	2	4	8	16	32	64	128	256

This means that every multiple of 64 is a separate subnet, the ranges 10.0.0.0 – 10.63.255.255 (only goes up to 10.63.255.255 because the next address, 10.64.0.0, is the network ID of the next subnet); 10.64.0.0 – 10.127.255.255; 10.128.0.0 – 10.191.255.255; and lastly, 10.192.0.0 – 10.255.255.255. We assign the first range to the marketing department, the second to the sales department and the third subnet to the engineering department. There is another subnet left, 10.192.0.0 – 10.255.255.254, which we have not assigned as we have no need for it.

To reinforce an earlier concept, what is another way to display our subnet mask of 255.192.0.0 that is much quicker to write? We can simply use /10 to signify our prefix length. This indicates that we are using 10 bits for the network, 8 for the network ID and 2 for the subnet ID. So for the IP address of 10.0.0.1 it would look like 10.0.0.1 /10.

Figure 3.4: Breakdown of the Sales Subnet Address

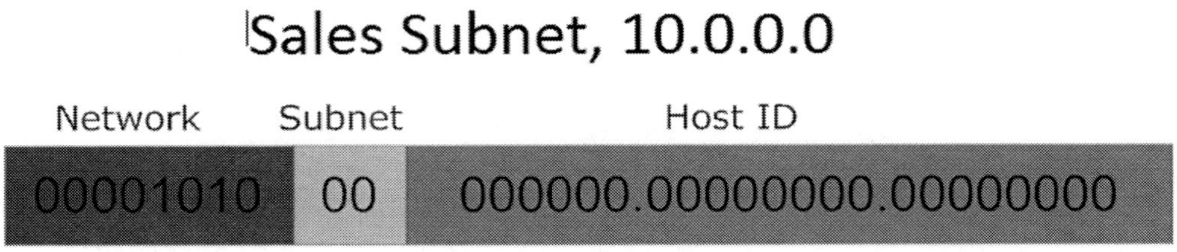

Sales Subnet, 10.0.0.0

Network	Subnet	Host ID
00001010	00	000000.00000000.00000000

The first subnet, as previously mentioned, ranges from 10.0.0.0 – 10.63.255.255. Notice in Figure 3.4 that the subnet portion is also included as part of the subnet mask because it represents the subnetwork. The minimal value is 0 because with all the bits in the second octet off they equal 0. The reason the second octet in our example only goes up to 63 is because that is the maximum total value of the 6 least significant bits (32 + 16 + 8 + 4 +2 + 1 = 63); as the two most significant bits of the second octet are 00. Don't worry if you have to reread that a few times as many people do.

Again referring to Figure 3.4, this leaves us with 22 bits for the host portion of the address. Per our formula, (multiply two by itself <number of host bits> times and subtract the network and broadcast addresses (which is the minus 2). 2 multiplied by itself 22 times is 4,194,304; minus 2, the answer would be 4,194,302. Hence, we have 4,194,302 valid host addresses.

Network: 10.0.0.0

Subnet mask: 255.192.0.0

Address range: 10.0.0.0 – 10.63.255.255

Network address: 10.0.0.0

Broadcast address: 10.63.255.255

Valid host range: 10.0.0.1 – 10.63.255.254

Figure 3.5: Breakdown of the Marketing Subnet Address

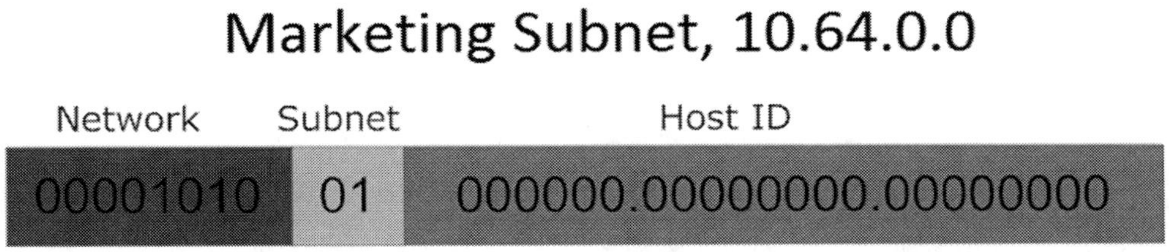

Marketing Subnet, 10.64.0.0

Network	Subnet	Host ID
00001010	01	000000.00000000.00000000

In Figure 3.5 the two most significant bits of the second octet are 0 and 1 respectively. The minimal value is 64 because the binary value of the second octet with all the host bits off is 64. The maximum value with all the host bits on is 127 (64 + 32 + 16 + 8 + 4 + 2 + 1).

Network: 10.64.0.0

Subnet mask: 255.192.0.0

Address range: 10.64.0.0 – 10.127.255.255

Network address: 10.64.0.0

Broadcast address: 10.127.255.255

Valid host range: 10.64.0.1 – 10.127.255.254

Figure 3.6: Breakdown of the Engineering Subnet Address

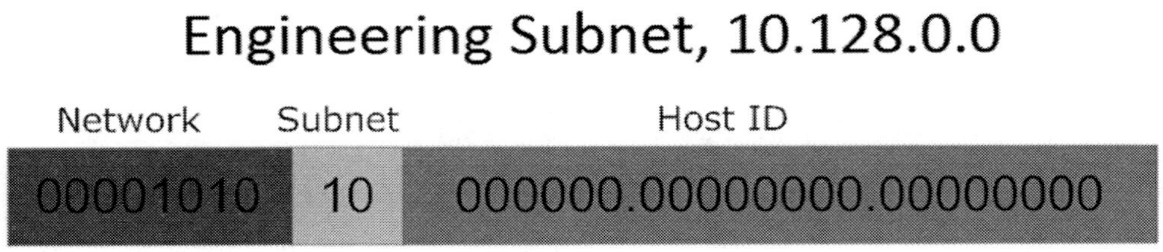

Engineering Subnet, 10.128.0.0

Network	Subnet	Host ID
00001010	10	000000.00000000.00000000

Figure 3.6 displays that in the third subnet, the first and second bits of the second octets are set to 1 and 0 respectively, making the minimum value of the second octet 128. Because if all bits are set to 0 the first two bits of the second octet remain 1 and 0 as they're part of the network ID which cannot be manipulated. So with all the host bits off we have a minimal value of 128. With all the host bits on, we have a string 10111111. Hence, the maximum value would be 191. Is it starting to get easy now?

Network: 10.128.0.0

Subnet mask: 255.192.0.0

Address range: 10.128.0.0 – 10.191.255.255

Network address: 10.128.0.0

Broadcast address: 10.191.255.255

Valid host range: 10.128.0.1 – 10.191.255.254

Figure 3.7: Breakdown of the 4th Subnet Address

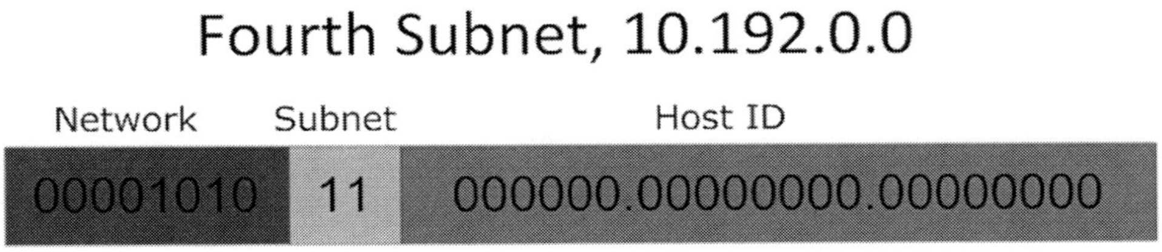

Fourth Subnet, 10.192.0.0

Network	Subnet	Host ID
00001010	11	000000.00000000.00000000

Finally in Figure 3.7, we see that the fourth subnet ranges from a minimum of 192 (binary string minimum of 11000000) and a maximum of 255 (binary string 11111111).

Network: 10.192.0.0

Subnet mask: 255.192.0.0

Address range: 10.192.0.0 – 10.255.255.255

Network address: 10.192.0.0

Broadcast address: 10.255.255.255

Valid host range: 10.192.0.1 – 10.255.255.254

Finding the Number of Valid Hosts

Now wait a minute, these subnets provide us with 4,194,302 valid host addresses! That's a lot of addresses we won't even get to use! So as a refresher, what is an easy way to determine the number of hosts? It's the number of our host bits to the 2nd power minus 2. Don't worry; we will give you an opportunity to do some practice questions determining the number of valid hosts based upon our subnet mask. But for now, if we want to save address space, we will use a longer subnet mask to have less host bits. Consider the following requirements using Figure 3.8 as a guide.

Figure 3.8: Required Number of Hosts

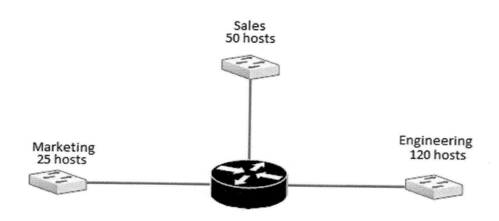

This is where the concept of VLSM (Variable Length Subnet Masks) comes in again. We need to find the subnet masks to use for 25, 50 and 120 hosts. Using our formula, we determine the number of host bits we need per subnet, and assign the remaining bits to the subnet portion. Effectively setting the network and host portions.

Generally as a best practice, we assign subnets to the network with the largest number of required hosts first and then work our way down to the subnets that require the least number of hosts. This will make your life a lot easier and is also easier to manage.

Starting with the Engineering department first, which needs 120 host assignable addresses. We calculate out how many host bits we need to fulfill their requirements. We will use Figure 3.9 as our power of 2 chart.

Figure 3.9: Power of 2 Chart

Bits	14	13	12	11	10	9	8	7	6	5	4	3	2	1
Hosts	16384	8192	4096	2048	1024	512	256	**128**	64	32	16	8	4	2

← ————————————————————————————— As you can see, the number of hosts doubles towards the left

Notice the hosts' value starts at **2** and increases as you move to the left. When using this table, you need to keep in mind that this chart provides to you the number of host IDs that can exist. The question on the exam will state you need 120 **valid** hosts. You need to make sure when using this chart you subtract two from our number to determine the number of valid hosts. Why? Well, we need to account a host ID for each the subnet ID and the subnet broadcast ID.

So back to our example of needing 120 valid hosts. So starting at the first bit with a value of 2, this is insufficient for our requirement so we move a bit to the left doubling the value to 4. This is still insufficient. Moving to the third bit from the right gives us 8 addresses, that's only sufficient for six valid hosts to communicate, our requirements state 120, 50 and 25. Considering that the number of hosts we need to satisfy is 120 for the engineering network, we can skip 8, 16, 32 and 64 (third, fourth, fifth and sixth bits from the right respectively) then we get to the seventh bit which gives us 128 addresses. It is the only logical choice as none of the previous would provide the sufficient number of addresses. This provides us with 128 addresses; actually with 126 valid ones as we need to subtract two for the network and broadcast addresses. Don't let that escape you as that is something they will try to trick you with on the exam.

What does this mean? Well, we have determined that we need 7 host bits to assign 120 addresses. To conserve address space and allow these addresses to be assigned to a network that actually needs the addresses. We only assign the bits we need to the host portion and assign the rest to the network portion, meaning that we will use a subnet mask inclusive of all bits except the bits we need for host addressing. In this case, we need 7 bits. So we use a subnet mask that covers the remaining 25 bits; which would be 11111111.11111111.11111111.10000000. As you can see, we left exactly 7 bits for host addressing. The subnet mask, when converted to decimal is 255.255.255.128 or /25 in CIDR notation (number of subnet bits) indicates the network portion.

Therefore, to fulfill the requirement for the engineering department, we use a mask of 255.255.255.128 to the 10.0.0.0 network; which makes the last octet the only one that can be used for host addressing.

Since the eighth bit is part of the network address as indicated by the subnet mask, we cannot use it for host addressing. Thus, the address range depends on the values of the seventh, sixth, fifth, fourth, third, second and first bits. If all of those host bits are off, they will equal zero. Thus making the minimum host ID 10.0.0.0 (all bits are 0) as shown in Figure 3.10.

Figure 3.10: Binary – Minimum Subnet Host ID Value

Bits	Eighth (leftmost)bit	Seventh bit	Sixth bit	Fifth bit	Fourth bit	Third bit	Second bit	First(rightmost) bit
4th octet	0	**0**	**0**	**0**	**0**	**0**	**0**	**0**
Value	128	64	32	16	8	4	2	1
Subnet Octet Value	128	192	224	240	248	252	254	255

The maximum as shown in Figure 3.11 would be 10.0.0.127. Hence the range 10.0.0.0 – 10.0.0.127 with all host IDs on.

Figure 3.11: Binary – Maximum Subnet Host ID Value

Bits	Eighth (leftmost)bit	Seventh bit	Sixth bit	Fifth bit	Fourth bit	Third bit	Second bit	First(rightmost) bit
4th octet	0	1	1	1	1	1	1	1
Value	128	64	32	16	8	4	2	1
Subnet Octet Value	128	192	224	240	248	252	254	255

So that gives us a network address of 10.0.0.0 and a broadcast address of 10.0.0.127 with all the addresses in between as **valid** host addresses.

Network: 10.0.0.0

Subnet mask: 255.255.255.128

Address range: 10.0.0.0 – 10.0.0.127

Network address: 10.0.0.0

Broadcast address: 10.0.0.127

Valid host range: 10.0.0.1 – 10.0.0.126

The sales department requires 50 hosts. So we use our table in Figure 3.12 to find out which binary values we are going to use.

Figure 3.12: Power of 2 Chart

Bits	14	13	12	11	10	9	8	7	**6**	5	4	3	2	1
Hosts	16384	8192	4096	2048	1024	512	256	128	**64**	32	16	8	4	2

We can either go with 32 which requires five bits and will yield an insufficient number of addresses or pick 64 which requires six bits and have an excess of 12 addresses(technically 10 once we subtract our 2, but the point for now is we have enough and some extras). So we will use 6 bits for the host portion as shown in Figure 3.13 and assign the remaining 26 bits to the network partition by using a /26 subnet mask or 255.255.255.192.

Figure 3.13: Binary – Marketing Host Portion of the Address

Bits	Eighth (leftmost)bit	Seventh bit	Sixth bit	Fifth bit	Fourth bit	Third bit	Second bit	First(rightmost) bit
4th octet	1	1	0	0	0	0	0	0
Value	128	64	32	16	8	4	2	1
Subnet Octet Value	128	192	224	240	248	252	254	255

The 6 remaining bits are going to be used for host addressing.

11111111.11111111.11111111.11**000000**

The first 26 bits are going to be used in the network portion. Hence, they will be represented by a subnet mask of all 1s. Making the mask 26 1s long and just the host portion will be left as 0s.

Since the engineering department got the range 10.0.0.0 – 10.0.0.127, we will use the range 10.0.0.128 – 10.0.0.191 to keep our addressing scheme contiguous. Now wait a second here. The first range was 128, why is this range 64? This is the very essence of VLSM. For the Engineering group by using a subnet mask of /25 I was able to have a range of 128 and for the Sales group by having a subnet mask of /26 I am able to have a range of 64 hosts which conserves space!

Network: 10.0.0.128

Subnet mask: 255.255.255.192

Address range: 10.0.0.128 – 10.0.0.191

Network address: 10.0.0.128

Broadcast address: 10.0.0.191

Valid host range: 10.0.0.129 – 10.0.0.190

You can think of subnet masks as a way to "suppress" bits from being used in the host portion to avoid wasting addresses. We suppressed fewer bits in the sales department (shorter mask) as we needed more host bits to satisfy our requirements. Remember, it is done at our own discretion.

The marketing department needs 25 addresses. Again, we use our power of 2 chart in Figure 3.14 and identify the number of bits we need.

Figure 3.14: Power of 2 Chart

Bits	14	13	12	11	10	9	8	7	6	5	4	3	2	1
Hosts	16384	8192	4096	2048	1024	512	256	128	64	32	16	8	4	2

5 bits, (2 multiplied by itself 5 times is 32) are sufficient. So will assign five bits for the host portion as shown in Figure 3.15 and that gives us 30 valid host addresses. So we use a /27 mask. Continuing with the earlier addressing scheme, we will start from 10.0.0.192, and go up to 10.0.0.223.

Figure 3.15: Binary – Host Portion of the Address

Bits	Eighth (leftmost)bit	Seventh bit	Sixth bit	Fifth bit	Fourth bit	Third bit	Second bit	First(rightmost) bit
4th octet	1	1	1	0	0	0	0	0
Value	128	64	32	16	8	4	2	1
Subnet Octet Value	128	192	224	240	248	252	254	255

Network: 10.0.0.192

Subnet mask: 255.255.255.224

Address range: 10.0.0.192 – 10.0.0.223

Network address: 10.0.0.192

Broadcast address: 10.0.0.223

Valid host range: 10.0.0.193 – 10.0.0.222

Our final addressing scheme would look as follows in Figure 3.16:

Figure 3.16: Final Subnetted Address Scheme

By understanding the concepts we discussed above, you should be able to determine what the network ID is, the first valid host, the last valid host and the broadcast ID for the subnet. A neat little tip for this is as follows:

Network ID: Always Even
First Valid Host: Always Odd
Last Valid Host: Always Even
Broadcast ID: Always Odd

Now that we have reviewed the theory behind how to figure out the subnet mask we would use when presented a requirement for a certain number of hosts on a subnet, we will do our customary practice to cement the concept for us.

First we are going to give you four network IDs and subnet masks and you will tell us the number of valid hosts that exist.

The Block Method – Subnetting For Speed!

Now that you understand the foundations of how and why we subnet, decimal to binary conversion (and vice versa), we are going to look at how we can perform these calculations with speed! This will hopefully free up some of your exam time for other areas such as scenario based questions.

As previously mentioned each octet has 256 possible values (0-255), each with a maximum value of 255. We are now going to look at examples of how we can use this information to perform subnetting quickly. You should think of the different network subnets as blocks. With each block denoting a specific size for the subnet. Let's try an example with the information below:

IP address: 10.1.1.1
Subnet Mask: 255.255.**248**.0

Looking at the above example we must first identify at which octet the network is being subnetted. We do this by comparing the IP Address against the subnet mask, by this point we know that 255 is the maximum value of an octet so we can ignore any IP address octet that matches a subnet octet with a value of 255. In the above example we can see the 3rd octet is where the subnetting is taking place as it has a value of less than 255. Now that we have this information to work out the block size all we do is take the number used within that octet (in this case 248), and subtract it from 256 (which is the number of possible values within an octet). Using the example above 256-248 = 8, therefore 8 is our block size. We now know that the IP address in our example is part of a network that is divided into 8 block increments. The lowest address in the block will always be the network address whereas the highest address will be the broadcast, and all interim addresses are valid assignable addresses.

Now if we look again at the IP Address 10.1.1.1 we know that this host resides in a network address of 10.1.0.0 and a broadcast address of 10.1.7.255. The next subnet (assuming the same subnet mask is utilized) would have a network address of 10.1.8.0 and a broadcast address of 10.1.15.255.

Now let's try this again

IP address: 10.1.45.1
Subnet Mask: 255.255.**224**.0

256-224 = 32, so we know that we are operating in 32 block increments, so if we look at the example above we can now see that the matching IP address octet is greater than 32 but less than 64 so we know that the address lives within the 10.1.32.0 network as the first network only contains addresses 10.1.0.1 – 10.1.31.255. So we know the host resides in a subnet with a network address of 10.1.32.0 and a broadcast address of 10.1.63.255.

Now there is actually another way that this same concept can be represented and that is using CIDR. If you recall from the beginning of chapter three, CIDR is how we can quickly display the number of bits that are used in a subnet mask. This value can span from /8 (255.0.0.0) to a theoretical 32 (255.255.255.255 not that you would actually ever use this value in application and the more practical value of /30 or 255.255.255.252 which is great for point-to-point links). By now we should understand that each natural class boundary is every 8 bits. So our natural class boundaries are /8 (255.0.0.0), /16 (255.255.0.0), /24 (255.255.255.0), and /32 (255.255.255.255 as mentioned before in practical use this generally will not exceed /30 but you will see why in a second why we use /32). So again to summarize, our boundaries are 8, 16, 24 and 32.

So let's look at our first example from our subnetting for speed and use the same numbers. Our IP address is 10.1.1.1 and our subnet mask is /21. You simply ask yourself what is the next boundary number higher than the current CIDR I have? In this case my mask is /21 so my next higher boundary number would be 24.

We will subtract our mask from our boundary number. So 24 – 21 = 3. Now $2^3 = 8$ which gives us our block size. Does this match our initial example? Yes! So I am not going to tell you what method to use. I am just going to provide you with the knowledge and power to pick the one you like best. However if you ask me, I like the first method presented the best as I think you can quickly convert from a CIDR subnet mask format to displaying it in dotted decimal format.

Let's do one more example using our second example above. Our ip address is 10.1.45.1 and the subnet mask of /19. So again we are at the 24 boundary and 24 – 19 = 5. $2^5 = 32$ and thus would be our block size. Does that match our example above? Yes! When we do our review exercises, you can pick which method you would like to use and we will explain the answers for both methodologies.

Exam Tip – Cisco is notorious for trying to trip you up on the exam in the way they ask the questions. So make sure you take the extra time to read the question carefully. For example, they may ask you what IP address is **not** in the valid host range. So for answer A they will give you an answer that **is** in the valid host range. The student will many times jump and pick that answer and go on to the next question. Now they just got that question incorrect. I generally will make sure none of the other answers could possibly fit. If they do, I know I made a mistake and need to review the question again.

Exercise 3.4: Finding the Number of Valid Hosts

To solve the following exercises, you simply need to find out which bits of the address belong to the network part and which bits belong to the host part. You can use the 2 to power of the number of host bits minus 2 formula or apply the subnetted octet to the chart in Figure 3.14 (also don't forget the minus 2 here). You can even do a little of both!

1) 221.1.54.0 /26

Masks	1st Octet	2nd Octet	3rd Octet	4th Octet
Class C mask				
/26 Mask				

Number of valid hosts: _____

2) 130.54.0.0 /21

Masks	1st Octet	2nd Octet	3rd Octet	4th Octet
Class C mask				
/21 Mask				

Number of valid hosts: _____

3) 223.5.191.0 /30

Masks	1st Octet	2nd Octet	3rd Octet	4th Octet
Class C mask				
/30 Mask				

Number of valid hosts: _____

4) 14.0.0.0 /18

Masks	1st Octet	2nd Octet	3rd Octet	4th Octet
Class C mask				
/18 Mask				

Number of valid hosts: _____

Exercise 3.5: Choosing the Appropriate Subnet Mask

Below are four scenarios with a network ID and number of required hosts. Find the correct subnet mask to use without allowing for too many hosts.

1) 127 hosts are required from the 10.0.0.0 network.

Bits order	Eighth (leftmost)bit	Seventh bit	Sixth bit	Fifth bit	Fourth bit	Third bit	Second bit	First(rightmost) bit
Bits								
Value	128	64	32	16	8	4	2	1
Subnet Octet Value	128	192	224	240	248	252	254	255

Subnet Mask _____

2) 18 hosts are required from the 131.15.0.0 network.

Bit order	Eighth (leftmost)bit	Seventh bit	Sixth bit	Fifth bit	Fourth bit	Third bit	Second bit	First(rightmost) bit
Bits								
Value	128	64	32	16	8	4	2	1
Subnet Octet Value	128	192	224	240	248	252	254	255

Subnet Mask _____

3) 194 hosts are required from the 219.151.15.0 network.

Bit order	Ninth bit (start of the 3rd octet)	Eighth bit (end of the 4th octet)	Seventh bit	Sixth bit	Fifth bit	Fourth bit	Third bit	Second bit	First(rightmost) bit
Bits									
Value	256	128	64	32	16	8	4	2	1
Subnet Octet Value	255	128	192	224	240	248	252	254	255

Subnet Mask _____

4) 549 hosts are required from the 14.0.0.0 network.

Bit order	Eighth (leftmost)bit	Seventh bit	Sixth bit	Fifth bit	Fourth bit	Third bit	Second bit	First(rightmost) bit
Bits								
Value	128	64	32	16	8	4	2	1
Subnet Octet Value	128	192	224	240	248	252	254	255

Subnet Mask _____

Exercise 3.6: Finding the Network ID, Broadcast Address, Address and Valid Address Ranges

Third we are going to have you determine the network ID, the broadcast address and valid IP address ranges for a given address based upon the knowledge you have acquired.

1) 22.21.48.120/26

Bits	1st Octet	2nd Octet	3rd Octet	4th Octet
IP Address				
Subnet Mask				
Network ID				

Network: _____

Subnet mask: _____

Address range: _____

Network address: _____

Broadcast address: _____

Valid host range: _____

2) 150.8.8.65/23

Bits	1st Octet	2nd Octet	3rd Octet	4th Octet
IP Address				
Subnet Mask				
Network ID				

Network: _____

Subnet mask: _____

Address range: _____

Network address: _____

Broadcast address: _____

Valid host range: _____

3) 195.25.30.154/27

Bits	1st Octet	2nd Octet	3rd Octet	4th Octet
IP Address				
Subnet Mask				
Network ID				

Network: _____

Subnet mask: _____

Address range: _____

Network address: _____

Broadcast address: _____

Valid host range: _____

Exercise 3.7: Subnetting for Speed!

In this section we are just going for speed. I will assume you know how to determine the network ID, broadcast address, total address range, and the valid host range once we determine the block size.

1) 10.34.0.0 mask 255.240.0.0

_____

2) 192.168.10.0 /28

_____

3) 182.16.0.0 mask 255.255.128.0

_____

4) 172.16.0.0 /29

_____

5) 21.0.0.0 mask 255.255.192.0

_____

Route Summarization

Routers use routing tables to keep track of which networks are reachable out which interfaces. This can cause memory and CPU problems if we have large routing tables as lookups take longer due to searching the routing table for a match to send traffic out of.

To demonstrate this, let us assume that our company's router is connected to another router as shown in Figure 3.16.

Figure 3.16

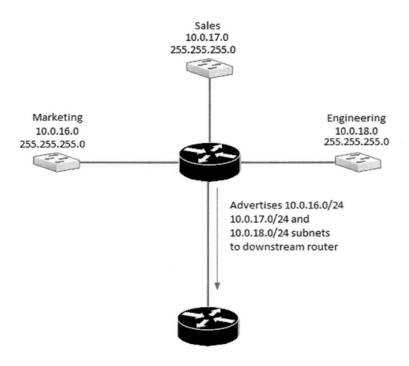

Routers use a "routing protocol" to exchange route tables that are directly connected and learned from other routers. Here, the center upstream router sends to the downstream router route information of the networks that are directly connected to it. Thus enabling the downstream router to reach those subnets as well.

What if we have hundreds of routers connected and thousands of routes? Well, as previously mentioned that would overwhelm the router and probably cause it to crash for most implementations and small companies. Are there routers out there that can handle let's say 300,000 routes? Yes, they are Internet backbone routers and are extremely expensive. So part of what you will need to do as a network engineer is size your routers appropriate to your environment.

If we have contiguous routes, we can summarize those routes instead of advertising specific routes. We can advertise summaries to those routes. In our case, we simply use a shorter prefix length that includes all of the more specific routes we want to advertise as in our example in Figure 3.17. So we will want to find a subnet mask which encompasses the range 10.0.16.0 – 10.0.18.255, which is the total range of addresses we have in the marketing, sales and engineering networks as previously shown in Figure 3.16.

So how do we find the summary route? We simply line up the three subnets in binary as shown in Figure 3.17. We then see which leading bits match between all the subnets (they are shown in bold below). We then turn all the matching bits to on to see the resultant summary route.

Figure 3.17

00001010.00000000.00010000.00000000 – Marketing Subnet

00001010.00000000.00010001.00000000 – Sales Subnet

00001010.00000000.00010010.00000000 – Engineering Subnet

11111111.11111111.11111100.00000000 – Summary Route

Count the number of bits that are on (there are 22 bits on) and we now have our mask! Pretty simple, huh?

Remember, the 4th octet is reserved for host usage in all networks and we cannot "unsupress" any subnet bits from it because there aren't any as they are host bits. So as you can see from our chart in Figure 3.18, with the least significant bit that is turned on (the third bit), we will advertise subnets in blocks of 4.

So if you refer to Figure 3.18, you will see this does give us another mathematical shortcut to figure out block sizes which can be summarized. Hopefully now you can see why we show the shortcut way and long binary way to figure out all these scenarios so you can really understand them and see the relationships.

Figure 3.18

Bits	Eighth (leftmost)bit	Seventh bit	Sixth bit	Fifth bit	Fourth bit	Third bit	Second bit	First(rightmost) bit
3th octet	1	1	1	1	1	1	0	0
Number of subnets per block	128	64	32	16	8	4	2	1

From our table in Figure 3.18 (which is for the third octet), we see that we can use all 6 bits for the summary range 10.0.16.0 – 10.0.19.255 (/22 mask). Then you can see how this applies to our topology example in Figure 3.19 which shows how the /22 summary advertisement will look.

Figure 3.19

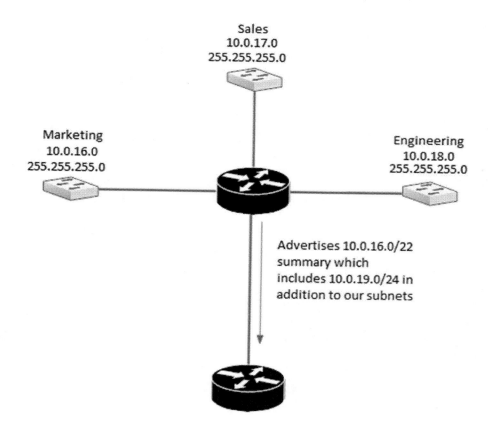

Something you may note is we are advertising a route for the 10.0.19.0 network and we don't actually have that sitting behind this router. That could potentially cause a problem and potentially keeping traffic from reaching its destination. There are various ways to address this problem, but they are outside the scope of this text.

So let's do three exercises of determining what would be the summary route and subnet mask advertised for the following scenarios (we did not include the full charts for these as they would be very large, you may want to do them on scrap paper).   The chart, although not required, will probably come in handy to check your work.

Exercise 3.8: Determining Correct Masks for Summarization

1) 10.0.48.0 to 10.0.63.0

Bits	Eighth (leftmost)bit	Seventh bit	Sixth bit	Fifth bit	Fourth bit	Third bit	Second bit	First(rightmost) bit
3th octet								
Number of subnets per block	128	64	32	16	8	4	2	1
Subnet Octet Value	128	192	224	240	248	252	254	255

Summary Route _____   Subnet Mask_____

2) 10.0.80.0 to 10.0.87.0

Bits	Eighth (leftmost)bit	Seventh bit	Sixth bit	Fifth bit	Fourth bit	Third bit	Second bit	First(rightmost) bit
3th octet								
Number of subnets per block	128	64	32	16	8	4	2	1
Subnet Octet Value	128	192	224	240	248	252	254	255

Summary Route _____   Subnet Mask_____

3) 10.0.160.0 to 10.0.191.0

Bits	Eighth (leftmost)bit	Seventh bit	Sixth bit	Fifth bit	Fourth bit	Third bit	Second bit	First(rightmost) bit
3th octet								
Number of subnets per block	128	64	32	16	8	4	2	1
Subnet Octet Value	128	192	224	240	248	252	254	255

Summary Route _____   Subnet Mask_____

Chapter 4

Meeting the Stated Design Requirements

Scenarios

In this chapter, we are going to put together everything we learned up until now by solving scenario based questions such as we will see on the exam. These are called the "Meeting the Stated Design Requirements" questions. So we will walk you through the logic on how to solve one and then we will provide you three practice examples for yourself.

We may see a sample such as the following:

Using network 205.75.10.0 you must develop a subnetting scheme that will allow for a minimum of 6 subnets and a maximum of 70 hosts. What is the best subnet mask to use?

When faced with a question like the one above we start by examining the stated requirements. In this case we are tasked with finding an appropriate mask that will create a minimum of 6 subnets and a maximum of 70 hosts, meaning that we can have subnets more than or equal to 6, because that is the minimum and we can go over it if we need to; and a maximum of 70 hosts on each subnet, meaning that we **must** have 70 or less hosts.

After examining the requirements, we need to find the number of bits we need to use to satisfy the requirements. We will start with the subnet portion as the host portion is more flexible than the subnet portion is.

The 205.75.10.0 network defaults to a class C mask, /24, we cannot modify neither the 1st, 2nd nor 3rd octets, so all of our subnetting will occur on the 4th octet.

Figure 4.1 Subnet Bit Chart

Bit order	Eighth bit	Seventh bit	Sixth bit	Fifth bit	Fourth bit	Third bit	Second bit	First bit
Bits	1	1	1	0	0	0	0	0
Subnets	2	4	8	16	32	64	128	256

As illustrated by Figure 4.1, since we need a minimum of 6 subnets, we can either go with 4 subnets or 8 subnets. 4 subnets would obviously not satisfy the requirement of having at least 6 subnets, so the number which makes the most sense is 8. While providing us with an additional two subnets going a little over the limit is not a problem in this case.

Figure 4.2 Host Bits Chart

Bits	Eighth (leftmost)bit	Seventh bit	Sixth bit	Fifth bit	Fourth bit	Third bit	Second bit	First(rightmost) bit
4th octet	1	1	0	0	0	0	0	0
Value	256	128	64	32	16	8	4	2

The number of hosts should be less than or equal to 70, which is between 128 and 64. 128 would be much higher than 70, so that is not a valid option as we are supposed to have a maximum of 70, the only logical choice is 64 as illustrated in Figure 4.2. This is really the same chart we used back in Figure 3.9 that tells us how many **host bits** we need to meet our number requirement. That means the other bits left over will be turned "on" as subnet bits. Back to our example, remember that we subtract 2 from the value (network ID and broadcast address) so we would have 62 valid host addresses. We are done, right? Nope. If we now compare the two masks they are different. So which one do we use or did we make a mistake? Well for either situation independent of the other, the resultant mask in Figure 4.1 of 255.255.255.224 or the resultant mask in Figure 4.2 of 255.255.255.192 would work. But we need to meet both requirements.

If we select the 255.255.255.192 mask in Figure 4.2, we then have 62 valid hosts but only 4 subnets. That does not meet the requirement. But if we select the 255.255.255.224 mask in Figure 4.1, then we have 8 subnets which meets the requirement of 6 and only 30 valid hosts per subnet. Not quite 64 which is the first number you might focus on under 70, but it now meets both requirements as 30 is less than 70. Don't let these trip you up on the exam. Take your time and read the questions as you do know the answers!

Exam Tip – Many times there will be multiple answers that could potentially fit for one requirement or the other. But there will only be one answer that meets both requirements. Always address the least flexibly requirement first and the rest should fall into place.

A guideline to follow is to always allocate the bits to the less flexible requirement. This means that you should always give bits to the portion that has explicit requirements. For example, specifying an exact value such as 62 hosts, or specifying a range such as 100-150 and so on. *Effectively leaving the flexible requirement, which is usually expressed in the terms less than or more than a certain value to be calculated last.*

Below is another example before getting to our practice exercises.

Allocate 100-150 subnets with at least 248 assignable host addresses from the 141.201.0.0/16 network.

Following the same steps we used in the previous example, we find the more restrictive requirement and satisfy it first. In this case, that would be the requirement for 100 to 150 subnets. To achieve this, we will need to subnet the third octet as the default mask of 255.255.0.0 only gives us one subnet. So we go back to our handy chart as shown in Figure 4.3 or use 2^7 to come up with our answer of 128 subnets and/or 7 bits resulting in a subnet mask of 255.255.254.0.

Figure 4.3 Third Octet Subnet Bits

Bits	Eighth (leftmost)bit	Seventh bit	Sixth bit	Fifth bit	Fourth bit	Third bit	Second bit	First(rightmost) bit
3rd octet	1	1	1	1	1	1	1	0
Value	2	4	8	16	32	64	128	256

In order to satisfy the second requirement which states we should have a number of hosts greater than or equal to 248. That is generally easy enough as if we leave the 4th octet all alone leaving a subnet mask of 255.255.255.0 we are left with 256 possible hosts minus 2 equals 254 hosts. But that then leaves us with 256 subnets. That is too many. So as we examine the two, we see that if we use our original subnet mask of 255.255.254.0 we will meet our restrictive requirement of between 100 and 150 subnets with 128 subnets and if we look at the chart in Figure 4.3 we will actually have 512 hosts minus 2 for a total of 510 as now the host portion extends into the 3rd octet. That is a lot more than the 248 we were seeking, but in reading the question closely, we just needed more than 248 hosts. It did not matter if we had 2 more or 2 million more hosts. We now met our requirement with 255.255.254.0.

Figure 4.3 Host Bits

Bit order	Ninth bit (start of the 3rd octet)	Eighth bit (end of the 4th octet)	Seventh bit	Sixth bit	Fifth bit	Fourth bit	Third bit	Second bit	First(rightmost) bit
Bits	0	0	0	0	0	0	0	0	0
Value	512	256	128	64	32	16	8	4	2

Now that we walked you through one example, we will provide you with three practice samples.

Exercise 4.1: Meeting the Stated Design with the Appropriate Subnet Mask

1) Using network 150.150.0.0, what subnet mask will result in 50 to 70 valid subnets that have at least 250 valid hosts per subnet?

Octets	1st Octet	2nd Octet	3rd Octet	4th Octet
Default Network Bits				
Subnet Bits				
Host Bits				

Resultant Subnet Mask _____ .

2) Your network number is 21.0.0.0. You need to have as many subnets as possible without exceeding 1000 subnets while at the same time having at least 500 hosts per subnet. What subnet mask would you use?

Octets	1st Octet	2nd Octet	3rd Octet	4th Octet
Default Network Bits				
Subnet Bits				
Host Bits				

Resultant Subnet Mask _____ .

3) Your network number is 199.20.6.0. You need to have at least 15 subnets. Each subnet should have between 5 and 12 hosts per subnet. Which subnet mask would you use?

Octets	1st Octet	2nd Octet	3rd Octet	4th Octet
Default Network Bits				
Subnet Bits				
Host Bits				

Resultant Subnet Mask _____ .

Chapter 5

Finite Address Space and NAT

Conserving IP Addresses

Some people may say the next section we are covering has absolutely nothing to do with subnetting. That can be argued a few different ways. But what I think we can all agree on is we subnet to save IP addresses. Accordingly due to this, we have developed other techniques to assist in saving IP addresses and we thought it would be prudent to touch on them in this section.

We previously mentioned that with 32 bits, we have a theoretical maximum of 4,294,967,295 IP addresses. It is "theoretical" because we have that many unique addresses assuming that each and every one of those addresses is a valid host address. However, that is unrealistic, as we have reserved addresses (0.0.0.0, 127.0.0.0), class D and class E addresses and network and broadcast IDs.

With the rapid expansion of the Internet, 4 billion addresses will not suffice because that is much less than the large demand of the world, not even considering that most users usually have more than one IP enabled device (laptops, desktops, workstations, smart phones, tablets, and emerging IP enabled household appliances). The finite address space is hindering growth and is limitation to a networked world.

To solve this problem IPv6 was proposed. Instead of 32 bits it uses 128 bits for addressing, allowing us to use roughly 340,000,000,000,000,000,000,000,000,000,000,000,000 addresses. That is quite a lot of addresses, ensuring that we will not run out in the foreseeable future. However, considering that the entire Internet is running IPv4 and some hosts already deployed are unable to support IPv6, an intermediary solution was required to alleviate the pressure and create a larger time span for migration to IPv6

RFC 1918 proposed the allocation of specific address ranges for host usage in local (private) internetworks which do not need public Internet connectivity; and hosts in local networks that do not require "immediate" (direct) connectivity to the Internet and can tolerate being connected through a public device.

What does this mean? Well, in cases hosts in enterprise networks do not need to access off-site resources (Internet), such as in basic client-server enterprise applications, the servers do not need Internet connectivity because the only clients they communicate with are on the local site. Therefore, assigning these hosts Internet usable (public) addresses would be wasteful as they are neither needed nor used. Also keep in mind that there are only so many IP addresses out there and available.

Figure 5.1: Intranet communication

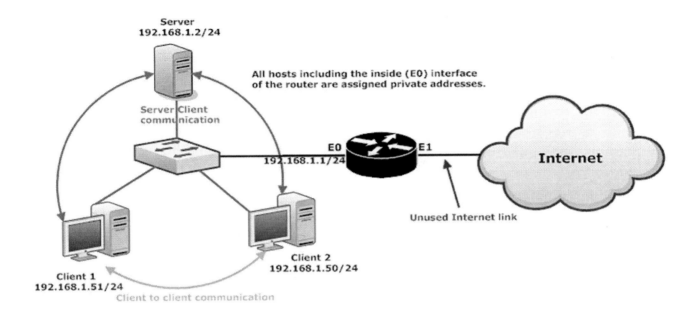

Per RFC 1918, we can assign hosts addresses in the 10.0.0.0/8, 172.16.0.0/12 and 192.168.0.0/16 ranges as shown in Figure 5.1. These ranges of addresses are especially allocated for this purpose. Considering that the same private address ranges are used across various organizations, enterprises and homes, they are not "routable" on the public Internet. If thousands of hosts have the same addresses, Internet routers would not be able to identify which host to send the traffic to.

But what about the enterprise hosts that need access to the Internet? In this case the Internet gateway (router) is assigned a public address on its Internet facing interface (E1 150.100.100.100 in Figure 5.2) which is used by the privately addressed hosts. This single address identifies the hosts by having the router set its public address as the source for all packets destined for the public Internet. This process of changing the source address from a private to a public address is done using a feature known as Network Address Translation (NAT).

Figure 5.2: Internet communication

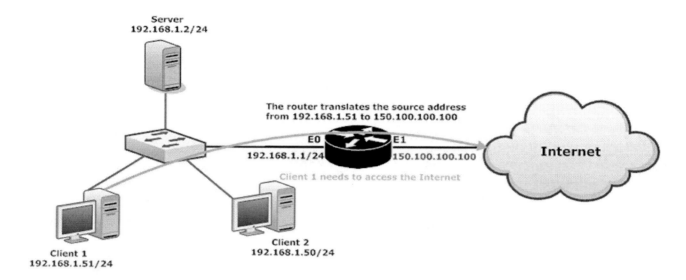

Network Address Translation (NAT)

NAT was devised as a method to alleviate the lack of IPv4 addresses while IPv6 matured and is actually deployed. However, it ended up greatly delaying the transition to IPv6 as the push to it was reduced because there was no longer an immediate need for IPv6. Basically NAT "translates" the private source address into a public one.

There are two ways to perform NAT. First is creating a one-to-one mapping between private and public addresses. For example, host 192.168.1.50 gets translated to 150.100.100.50; host 192.168.1.51 gets translated to 150.100.100.51, and so on. As you can see, this does not help solve the problem of the address shortage and defeats the purpose of NAT when deployed on a large scale. This application of NAT is usually used for servers and other devices that need to have an address that is reachable from the Internet, such as web and email servers. This is known as "static NAT" hinting that you statically create the private to public mappings of IP addresses. You can see an example of this in Figure 5.3.

Figure 5.3: NAT table

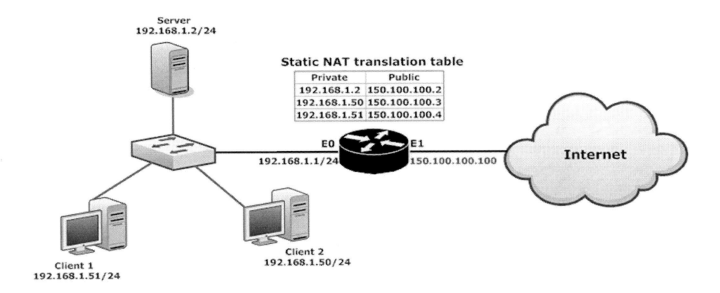

Dynamic NAT

The second method of deploying NAT is to create a pool of public addresses for NAT to use and dynamically allocate them to hosts as needed and return them to the pool once the hosts no longer need to access external resources (Internet) as shown in Figure 5.4. This is known as "dynamic NAT" it is very similar to static NAT in which every private address needs an equivalent public address. This again, does not help alleviate the lack of addresses.

Figure 5.4: Operation of Dynamic NAT

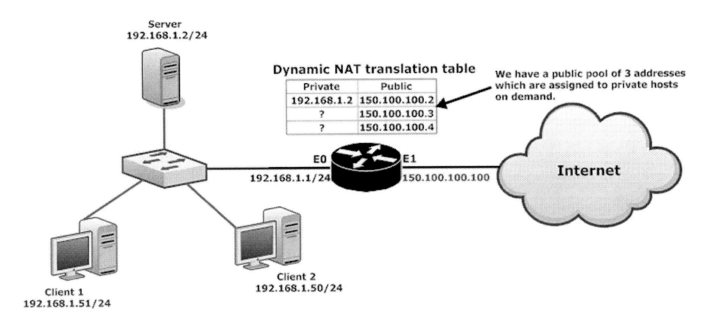

In our example in Figure 5.4, the only host currently accessing the Internet is the server, 192.168.1.2 and the remaining two public addresses are unallocated, meaning that they can be used by any host that needs to access the internet.

NAT Overload

What if we have more hosts than public addresses? This is where "NAT overload" comes in. Instead of associating a single private address with a public address, we associate or "translate" multiple private addresses to a single public address using the layer 4 (TCP and UDP) port numbers as shown in Figure 5.5.

Figure 5.5: NAT overload operation

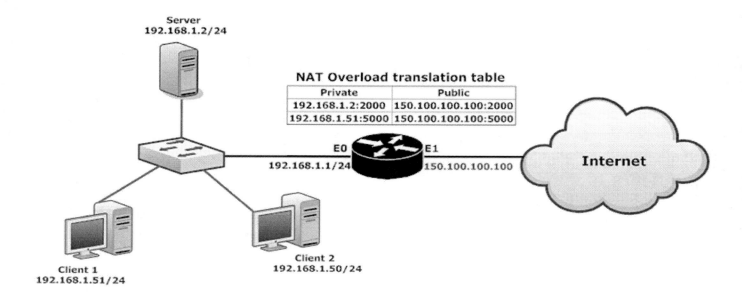

In this example, the port numbers are used to identify the host which originated the request. For instance, Client 1 sent a request out TCP port 5000 and instead of the router translating the private source address to a public one; it instead assigns a port to a single public address to identify the flow. In this case, the public address' port 2000 corresponds to 192.168.1.2 (server); and port 5000 corresponds to 192.168.1.51 (client 1). As you can see, this can go on and on, we are limited by the number of ports, which has a maximum of 65535 with several reserved ports. The reserved ports are not used by NAT by default, but it is possible to set up port forwarding for various applications which use well known port numbers. Note that NAT overload is also known as Port Address Translation (PAT) and is dynamic in nature. The port numbers are allocated dynamically and are withdrawn once hosts no longer need them.

So now you have a full understanding of not only how to subnet, but why we subnet!

Answer Key

Exercise 1.1: Binary to Decimal Practice

1) 11100111.00110010.00101101.11111011

Bits	Eighth (leftmost)	Seventh bit	Sixth bit	Fifth bit	Fourth bit	Third bit	Second bit	First bit	Total value
1st octet	1	1	1	0	0	1	1	1	**231**
2nd octet	0	0	1	1	0	0	1	0	**50**
3rd octet	0	0	1	0	1	1	0	1	**45**
4th octet	1	1	1	1	1	0	1	1	**251**
Value	128	64	32	16	8	4	2	1	

The correct answer is 231.50.45.251.

2) 00001110.01101110.00011001.00101010

Bits	Eighth (leftmost)	Seventh bit	Sixth bit	Fifth bit	Fourth bit	Third bit	Second bit	First bit	Total value
1st octet	0	0	0	0	1	1	1	0	**14**
2nd octet	0	1	1	0	1	1	1	0	**110**
3rd octet	0	0	0	1	1	0	0	1	**25**
4th octet	0	0	1	0	1	0	1	0	**42**
Value	128	64	32	16	8	4	2	1	

The correct answer is 14.100.25.42.

3) 11011101.00000001.11011101.01110110

Bits	Eighth (leftmost)	Seventh bit	Sixth bit	Fifth bit	Fourth bit	Third bit	Second bit	First bit	Total value
1st octet	1	1	0	1	1	1	0	1	**221**
2nd octet	0	0	0	0	0	0	0	1	**1**
3rd octet	1	1	0	1	1	1	0	1	**221**
4th octet	0	1	1	1	0	1	1	0	**118**
Value	128	64	32	16	8	4	2	1	

The correct answer is 221.1.221.118

4) 01010000.11111101.11110100.00010110

Bits	Eighth (leftmost)	Seventh bit	Sixth bit	Fifth bit	Fourth bit	Third bit	Second bit	First bit	Total value
1st octet	0	1	0	1	0	0	0	0	**80**
2nd octet	1	1	1	1	1	1	0	1	**253**
3rd octet	1	1	1	1	0	1	0	0	**244**
4th octet	0	0	0	1	0	1	1	0	**22**
Value	128	64	32	16	8	4	2	1	

The correct answer is 80.253.244.22.

5) 00000011.11111001.00110011.00111101

Bits	Eighth (leftmost)	Seventh bit	Sixth bit	Fifth bit	Fourth bit	Third bit	Second bit	First bit	Total value
1st octet	0	0	0	0	0	0	1	1	**3**
2nd octet	1	1	1	1	1	0	0	1	**249**
3rd octet	0	0	1	1	0	0	1	1	**51**
4th octet	0	0	1	1	1	1	0	1	**61**
Value	128	64	32	16	8	4	2	1	

The correct answer is 3.249.51.61.

6) 11000110.01011110.01111111.11111110

Bits	Eighth (leftmost)	Seventh bit	Sixth bit	Fifth bit	Fourth bit	Third bit	Second bit	First bit	Total value
1st octet	1	1	0	0	0	1	1	0	**198**
2nd octet	0	1	0	1	1	1	1	0	**94**
3rd octet	0	1	1	1	1	1	1	1	**127**
4th octet	1	1	1	1	1	1	1	0	**254**
Value	128	64	32	16	8	4	2	1	

The correct answer is 198.94.127.254.

7) 11111000.00000000.00000010.01100101

Bits	Eighth (leftmost)	Seventh bit	Sixth bit	Fifth bit	Fourth bit	Third bit	Second bit	First bit	Total value
1st octet	1	1	1	1	1	0	0	0	**248**
2nd octet	0	0	0	0	0	0	0	0	**0**
3rd octet	0	0	0	0	0	0	1	0	**2**
4th octet	0	1	1	0	0	1	0	1	**101**
Value	128	64	32	16	8	4	2	1	

The correct answer is 248.0.2.101.

8) 01000101.11111111.01011100.01111110

Bits	Eighth (leftmost)	Seventh bit	Sixth bit	Fifth bit	Fourth bit	Third bit	Second bit	First bit	Total value
1st octet	0	1	0	0	0	1	0	1	**69**
2nd octet	1	1	1	1	1	1	1	1	**255**
3rd octet	0	1	0	1	1	1	0	0	**92**
4th octet	0	1	1	1	1	1	1	0	**126**
Value	128	64	32	16	8	4	2	1	

The correct answer is 69.255.92.126.

9) 00000100.00001000.00000100.11111010

Bits	Eighth (leftmost)	Seventh bit	Sixth bit	Fifth bit	Fourth bit	Third bit	Second bit	First bit	Total value
1st octet	0	0	0	0	0	1	0	0	**4**
2nd octet	0	0	0	0	1	0	0	0	**8**
3rd octet	0	0	0	0	0	1	0	0	**4**
4th octet	1	1	1	1	1	0	1	0	**250**
Value	128	64	32	16	8	4	2	1	

The correct answer is 4.8.4.250.

10) 11011111.01110011.01000001.00101011

Bits	Eighth (leftmost)	Seventh bit	Sixth bit	Fifth bit	Fourth bit	Third bit	Second bit	First bit	Total value
1st octet	1	1	0	1	1	1	1	1	**223**
2nd octet	0	1	1	1	0	0	1	1	**115**
3rd octet	0	1	0	0	0	0	0	1	**65**
4th octet	0	0	1	0	1	0	1	1	**43**
Value	128	64	32	16	8	4	2	1	

The correct answer is 223.115.65.43.

Exercise 1.2 Decimal to Binary Practice

You may or may not have noticed. But the decimal to binary practice was the exact same numbers as the binary to decimal practice. So instead of reprinting all the same answers, you can just review the answers above.

Exercise 2.1: Determining the Network ID ANDing Practice

We simply align the IP address bits with the subnet mask bits to find out which part the bits belong to, the network part (including the subnet part) or the host part to find the correct answer in this exercise.

1)IP Address 21.153.24.169 with a subnet mask of 255.0.0.0

Bits	1st Octet	2nd Octet	3rd Octet	4th Octet
IP Address	00010101	10011001	00011000	10101001
Subnet Mask	11111111	00000000	00000000	00000000
Network ID	00010101	00000000	00000000	00000000

The IP address 21.153.24.168 is on the 21.0.0.0 network.

2) IP Address 131.1.54.69 with a subnet mask of 255.255.0.0

Bits	1st Octet	2nd Octet	3rd Octet	4th Octet
IP Address	10000011	00000001	00110110	01000101
Subnet Mask	11111111	11111111	00000000	00000000
AND answer	10000011	00000001	00000000	00000000

The IP address 131.1.54.69 is on the 131.1.0.0 network.

3) IP Address 239.244.1.16 with a subnet mask of 255.255.255.0

Bits	1st Octet	2nd Octet	3rd Octet	4th Octet
IP Address	11101111	11110100	00000001	00010000
Subnet Mask	11111111	11111111	11111111	00000000
AND answer	11101111	11110100	00000001	00000000

The IP address 239.244.1.16 is on the 239.244.1.0 network.

Exercise 2.2: Finding the Subnet the Address Resides On

1)IP Address 21.153.24.169 with a subnet mask of 255.240.0.0

Bits	1st Octet	2nd Octet	3rd Octet	4th Octet
IP Address	00010101	10011001	00011000	10101001
Subnet Mask	11111111	11110000	00000000	00000000
Network ID	00010101	10010000	00000000	00000000

The IP address 21.153.24.169 is on the 21.144.0.0 network.

2) IP Address 131.1.54.69 with a subnet mask of 255.255.192.0

Bits	1st Octet	2nd Octet	3rd Octet	4th Octet
IP Address	10000011	00000001	00110110	01000101
Subnet Mask	11111111	11111111	11000000	00000000
Network ID	10000011	00000001	00000000	00000000

The IP address 131.1.54.69 is on the 131.1.0.0 network.

3) IP Address 239.244.1.16 with a subnet mask of 255.255.255.192

Bits	1st Octet	2nd Octet	3rd Octet	4th Octet
IP Address	11101111	11110100	00000001	00010000
Subnet Mask	11111111	11111111	11111111	11000000
Network ID	11101111	11110100	00000001	00000000

The IP address 239.244.1.16 is on the 239.244.1.0 network.

Exercise 3.1: Determining the Number of Valid Subnets for Class A Addresses Practice

In the following three exercises we simply multiply two by itself by the number of excess mask bits to find the number of valid subnets.

1) 24.0.0.0 /14

Masks	1st Octet	2nd Octet	3rd Octet	4th Octet
Class A mask	11111111	00000000	00000000	00000000
/14 Mask	11111111	11111100	00000000	00000000

Number of excess mask bits = 6, 2 multiplied by itself 6 times is 64. So we have 64 valid subnets.

2) 2.0.0.0 /10

Masks	1st Octet	2nd Octet	3rd Octet	4th Octet
Class A mask	11111111	00000000	00000000	00000000
/10 Mask	11111111	11000000	00000000	00000000

Number of excess mask bits = 2, 2 multiplied by itself 2 times is 4. So we have 4 valid subnets.

3) 97.0.0.0 /17

Masks	1st Octet	2nd Octet	3rd Octet	4th Octet
Class A mask	11111111	00000000	00000000	00000000
/17 Mask	11111111	11111111	10000000	00000000

Number of excess mask bits = 9, 2 multiplied by itself 9 times is 512. So we have 512 valid subnets.

Exercise 3.2: Determining the Number of Valid Subnets for Class B Addresses Practice

1) 142.42.0.0 /25

Masks	1st Octet	2nd Octet	3rd Octet	4th Octet
Class B mask	11111111	11111111	00000000	00000000
/25 Mask	11111111	11111111	**11111111**	**1**0000000

Number of excess mask bits = 9, 2 multiplied by itself 9 times is 512. So we have 512 valid subnets.

2) 169.1.0.0 /19

Masks	1st Octet	2nd Octet	3rd Octet	4th Octet
Class B mask	11111111	11111111	00000000	00000000
/19 Mask	11111111	11111111	**111**00000	00000000

Number of excess mask bits = 3, 2 multiplied by itself 3 times is 8. So we have 8 valid subnets.

3) 131.31.0.0 /20

Masks	1st Octet	2nd Octet	3rd Octet	4th Octet
Class B mask	11111111	11111111	00000000	00000000
/20 Mask	11111111	11111111	**1111**0000	00000000

Number of excess mask bits = 4, 2 multiplied by itself 4 times is 16. So we have 16 valid subnets.

Exercise 3.3: Determining the Number of Valid Subnets for Class C Addresses Practice

1) 223.21.1.0 /30

Masks	1st Octet	2nd Octet	3rd Octet	4th Octet
Class C mask	11111111	11111111	11111111	00000000
/30 Mask	11111111	11111111	11111111	**111111**00

Number of excess mask bits = 6, 2 multiplied by itself 6 times is 64. So we have 64 valid subnets.

2) 210.21.9.0 /29

Masks	1st Octet	2nd Octet	3rd Octet	4th Octet
Class C mask	11111111	11111111	11111111	00000000
/29 Mask	11111111	11111111	11111111	**11111**000

Number of excess mask bits = 5, 2 multiplied by itself 5 times is 32. So we have 32 valid subnets.

3) 253.1.1.0 /26

Masks	1st Octet	2nd Octet	3rd Octet	4th Octet
Class C mask	11111111	11111111	11111111	00000000
/26 Mask	11111111	11111111	11111111	**11**000000

Number of excess mask bits = 2, 2 multiplied by itself 2 times is 4. So we have 4 valid subnets.

Exercise 3.4: Finding the Number of Valid Hosts

1) 221.1.54.0 /26

Mask	1st Octet	2nd Octet	3rd Octet	4th Octet
Class C mask	11111111	11111111	11111111	00000000
/26	11111111	11111111	11111111	11000000

Network: 221.1.54.0
Subnet mask: 255.255.255.192
Number of valid hosts: 62

We simply take the octets which contain 0s (host portion) and multiply 2 by itself <number of 0 or host bits> times and subtract 2. We **always** subtract 2 to allot for the network ID and broadcast address. So in this case we will multiply 2 by itself 6 times yielding 64 and subtract 2, resulting in 62 hosts.

2) 130.54.0.0 /21

Mask	1st Octet	2nd Octet	3rd Octet	4th Octet
Class C mask	11111111	11111111	11111111	00000000
/21	11111111	11111111	11111000	00000000

Network: 130.54.0.0
Subnet mask: 255.255.248.0
Number of valid hosts: 2046

Two multiplied by itself 11 times is 2048 minus 2, that is 2046 hosts.

3) 223.5.191.0 /30

Mask	1st Octet	2nd Octet	3rd Octet	4th Octet
Class C mask	11111111	11111111	11111111	00000000
/30	11111111	11111111	11111111	11111100

Network: 223.5.191.0
Subnet mask: 255.255.255.252
Number of valid hosts: 2

2 multiplied by itself twice is 4 minus 2, yielding 2 hosts.

4) 14.0.0.0 /18

Mask	1st Octet	2nd Octet	3rd Octet	4th Octet
/18	11111111	11111111	11000000	00000000

Network: 140.0.0.0
Subnet mask: 255.255.192.0
Number of valid hosts: 16,382

2 multiplied by itself 14 times is 16,384 minus 2 is 16,382 hosts.

Exercise 3.5: Choosing the Appropriate Subnet Mask

Find the correct subnet mask to use without allowing for too many hosts.

1) 127 hosts are required from the 10.0.0.0 network.

Bits order	Eighth (leftmost)bit	Seventh bit	Sixth bit	Fifth bit	Fourth bit	Third bit	Second bit	First(rightmost) bit
Bits	0	0	0	0	0	0	0	0
Value	128	64	32	16	8	4	2	1
Subnet Octet Value	128	192	224	240	248	252	254	255

Mask: 255.255.255.0 Remember, you always need to subtract two. If you picked 255.255.255.128 it provides 128 hosts minus 2 equals 126. That does not meet out requirement. So we just use the normal class C mask. Almost seems too simple, huh?

2) 18 hosts are required from the 131.15.0.0 network.

Bit order	Eighth (leftmost)bit	Seventh bit	Sixth bit	Fifth bit	Fourth bit	Third bit	Second bit	First(rightmost) bit
Bits	1	1	1	0	0	0	0	0
Value	128	64	32	16	8	4	2	1
Subnet Octet Value	128	192	224	240	248	252	254	255

Mask: 255.255.255.224

3) 194 hosts are required from the 219.151.15.0 network.

Bit order	Ninth bit (start of the 3rd bit)	Eighth bit (end of the 4th octet)	Seventh bit	Sixth bit	Fifth bit	Fourth bit	Third bit	Second bit	First(right most) bit
Bits	1	0	0	0	0	0	0	0	0
Value	256	128	64	32	16	8	4	2	1
Subnet Octet Value	255	128	192	224	240	248	252	254	255

Mask: 255.255.255.0 Hopefully this did not trip you up again. Note how in this example we show you the ninth bit.

4) 549 hosts are required from the 14.0.0.0 network.

Bit order	Tenth bit	Ninth bit (start of the 3rd bit)	Eighth bit (end of the 4th octet)	Seventh bit	Sixth bit	Fifth bit	Fourth bit	Third bit	Second bit	First(right most) bit
Bits	0	0	0	0	0	0	0	0	0	0

Value	512	256	128	64	32	16	8	4	2	1
Subnet Octet Value	254	255	128	192	224	240	248	252	254	255

Mask: 255.255.252.0 We kind of wanted to see what you would do if we did not give you a great big hint by putting the eleventh bit there. As you can see, even at ten bits out we only have 510 valid hosts. So we have to go to the eleventh bit which will give us 1024 hosts minus 2 equaling 1022 valid hosts.

Exercise 3.6: Finding the Network ID, Broadcast Address, Address and Valid Address Ranges

In this section we need to find the network ID, broadcast address, total address range, and the valid host range. The network ID is the first address in the range and the broadcast address is the last address in the range. To find these values, we simply align the network ID and subnet mask and find the correct network ID first, after which we calculate the ranges using the values specified by the subnet mask, i.e. find how many host bits exist, calculate their value, and add it to the network ID to find the total range. And lastly, to find the range of valid host assignable addresses, subtract 2, the network ID and the broadcast address. For a more through explanation, refer to the Finding number of Valid Hosts section of the How to and why we Subnet document.

1) 221.21.48.120 /26

Bits	1st Octet	2nd Octet	3rd Octet	4th Octet
IP Address	11011101	00010101	00110000	01111000
Subnet Mask	11111111	11111111	11111111	11000000
Network ID	11011101	00010101	00110000	01000000

Network: 221.21.48.64
Subnet mask: 255.255.255.192
Address range: 221.21.48.64 – 221.21.48.127
Network address: 221.21.48.64
Broadcast address: 221.21.48.127
Valid host range: 221.21.48.65 – 221.21.48.126

2) 150.8.8.65 /23

Bits	1st Octet	2nd Octet	3rd Octet	4th Octet
IP Address	10010110	00001000	00001000	01000001
Subnet Mask	11111111	11111111	11111110	00000000
Network ID	10010110	00001000	00001000	00000000

Network: 150.8.8.0
Subnet mask: 255.255.254.0
Address range: 150.8.8.0 – 150.8.9.255
Network address: 150.8.8.0
Broadcast address: 150.8.9.255
Valid host range: 150.8.8.1 – 150.8.9.254

3) 195.25.30.154 /27

Bits	1st Octet	2nd Octet	3rd Octet	4th Octet
IP Address	11000011	00011001	00011110	10011010
Subnet Mask	11111111	11111111	11111111	11100000
Network ID	11000011	00011001	00011110	10000000

Network: 195.25.30.128
Subnet mask: 255.255.255.224
Address range: 195.25.30.128 – 195.25.30.159
Network address: 195.25.30.128
 Broadcast address: 195.25.30.159
Valid host range: 195.25.30.129 – 195.25.30.158

Exercise 3.7: Subnetting for Speed!

In this section we are just going for speed. I will assume you know how to determine the network ID, broadcast address, total address range, and the valid host range once we determine the block size.

1) 10.34.0.0 mask 255.240.0.0

256 – 240 = a block size of 16 or mask 255.240.0.0 which is /12 and the next boundary number is 16. 16 – 12 = 4. 2^4 = 16.

2) 192.168.10.0 /28

Our next boundary number above 28 is 32. 32 – 28 = 4. 2^4 = 16. Additionally, /28 is 255.255.240.0. 256 – 240 = 16. So either way, you come up with a block size of 16 to help you figure out your various ranges.

3) 182.16.0.0 mask 255.255.128.0

256 – 128 = a block size of 128 or mask 255.255.128.0 which is /17 and the next boundary number is 24. 24 – 17 = 7. 2^7 = 128

4) 172.16.0.0 /29

Our next boundary number above 29 is 32. 32 – 29 = 3. 2^3 = 8. Additionally, /29 is 255.255.248.0. 256 – 248 = 8.

5) 21.0.0.0 mask 255.255.192.0

256 – 192 = a block size of 64 or mask 255.255.192.0 which is /18 and the next boundary number is 24. 24 – 18 = 6. 2^6 = 64.

Exercise 3.8: Determining Correct Masks for Summarization

1) 10.0.48.0 to 10.0.63.0

Bits	Eighth (leftmost)bit	Seventh bit	Sixth bit	Fifth bit	Fourth bit	Third bit	Second bit	First(rightmost) bit
3th octet	1	1	1	1	0	0	0	0
Number of subnets per block	128	64	32	16	8	4	2	1
Subnet Octet Value	128	192	224	240	248	252	254	255

Summary Route 10.0.48.0   Subnet Mask 255.255.240.0

2) 10.0.80.0 to 10.0.87.0

Bits	Eighth (leftmost)bit	Seventh bit	Sixth bit	Fifth bit	Fourth bit	Third bit	Second bit	First(rightmost) bit
3th octet	1	1	1	1	1	0	0	0
Number of subnets per block	128	64	32	16	8	4	2	1
Subnet Octet Value	128	192	224	240	248	252	254	255

Summary Route 10.0.80.0   Subnet Mask 255.255.248.0

3) 10.0.160.0 to 10.0.191.0

Bits	Eighth (leftmost)bit	Seventh bit	Sixth bit	Fifth bit	Fourth bit	Third bit	Second bit	First(rightmost) bit
3th octet	1	1	1	0	0	0	0	0
Number of subnets per block	128	64	32	16	8	4	2	1
Subnet Octet Value	128	192	224	240	248	252	254	255

Summary Route 10.0.160.0   Subnet Mask 255.255.224.0

Exercise 4.1: Meeting the Stated Design with the Appropriate Subnet Mask

1) Using network 150.150.0.0, what subnet mask will result in between 50 to 70 valid subnets that have at least 250 valid hosts per subnet?

Octets	1st Octet	2nd Octet	3rd Octet	4th Octet
Default mask	11111111	11111111	00000000	00000000
New mask	11111111	11111111	**11111100**	00000000

Using 6 bits for the subnet mask provides us with exactly 64 valid subnets. Since the question states 50 to 70, and with 64 being the only number inside that range, it is the correct choice. This leaves us with 12 host bits, resulting in $1024 - 2$ or 1022 valid host addresses, satisfying the requirements. Therefore, the final answer is 255.255.252.0

2) Your network number is 21.0.0.0. You need to have as many subnets as possible without exceeding 1000 subnets while at the same time having at least 500 hosts per subnet. What subnet mask would you use?

21.0.0.0 defaults to a /8 mask; leaving us with 24 bits to allocate as we please. Considering the requirements, we need less than or equal to 1,000 subnets. So we need to find out how many times we need to multiply 2 by itself to find the number of bits we should use for the subnet portion.

Octets	1st Octet	2nd Octet	3rd Octet	4th Octet
Default mask	11111111	00000000	00000000	00000000
New mask	11111111	**11111111**	10000000	00000000

Using 9 bits for the subnet mask in addition to the existing 8 bits (making it a /17). To understand why we chose 9 additional subnet bits, it is because our only logical choices were 10 bits-1024 as 2 multiplied by itself 10 times is 1024, and 9 bits-512. Since the question stated not to go above 1000, 1024 was out of the question.

Since we used 17 bits for the mask, we are left with 15 bits for host addressing, which yields 32,766 host addresses; more than enough to satisfy the requirements. Making the appropriate subnet mask 255.255.128.0

3) Your network number is 199.20.6.0. You need to have at least 15 subnets. Each subnet should have between 5 and 12 hosts per subnet. Which subnet mask would you use?

Octets	1st Octet	2nd Octet	3rd Octet	4th Octet
Default mask	11111111	11111111	11111111	00000000
New mask	11111111	11111111	11111111	**11111000**

We need at least 15 subnets and 5-12 hosts per subnet. Now some students will start off and address the number of subnets first saying, hey...this is easy. We will do 16 subnets as that is more than 15 and we then have a subnet mask of 255.255.255.240 and that gives us 14 hosts per subnet. The only problem with that is we actually have too many hosts per subnet. So if we go up a little more and have more subnets than is really required with 32 (but not exceeding the requirement) we will then see that our number of hosts drops down to 6 meeting the stated requirement. So how does that look like in binary? We need to make the number of subnets greater than or equal to 15, thus making it more flexible than the number of hosts required. So we will allocate bits for the host portion first and use the remaining bits as the subnet mask. So we will use 3 bits for the host portion($2^3-2=6$) and allocate the rest to the subnet portion; which is 255.255.255.248 in binary.